George Harley

Histological Demonstrations

A guide to the microscopical examination of the animal tissues in health and disease for the use of the medical and veterinary professions : being the substance of lectures delivered

George Harley

Histological Demonstrations
A guide to the microscopical examination of the animal tissues in health and disease for the use of the medical and veterinary professions : being the substance of lectures delivered

ISBN/EAN: 9783337075392

Printed in Europe, USA, Canada, Australia, Japan

Cover: Foto ©berggeist007 / pixelio.de

More available books at **www.hansebooks.com**

HISTOLOGICAL DEMONSTRATIONS.

LONDON
PRINTED BY SPOTTISWOODE AND CO.
NEW-STREET SQUARE

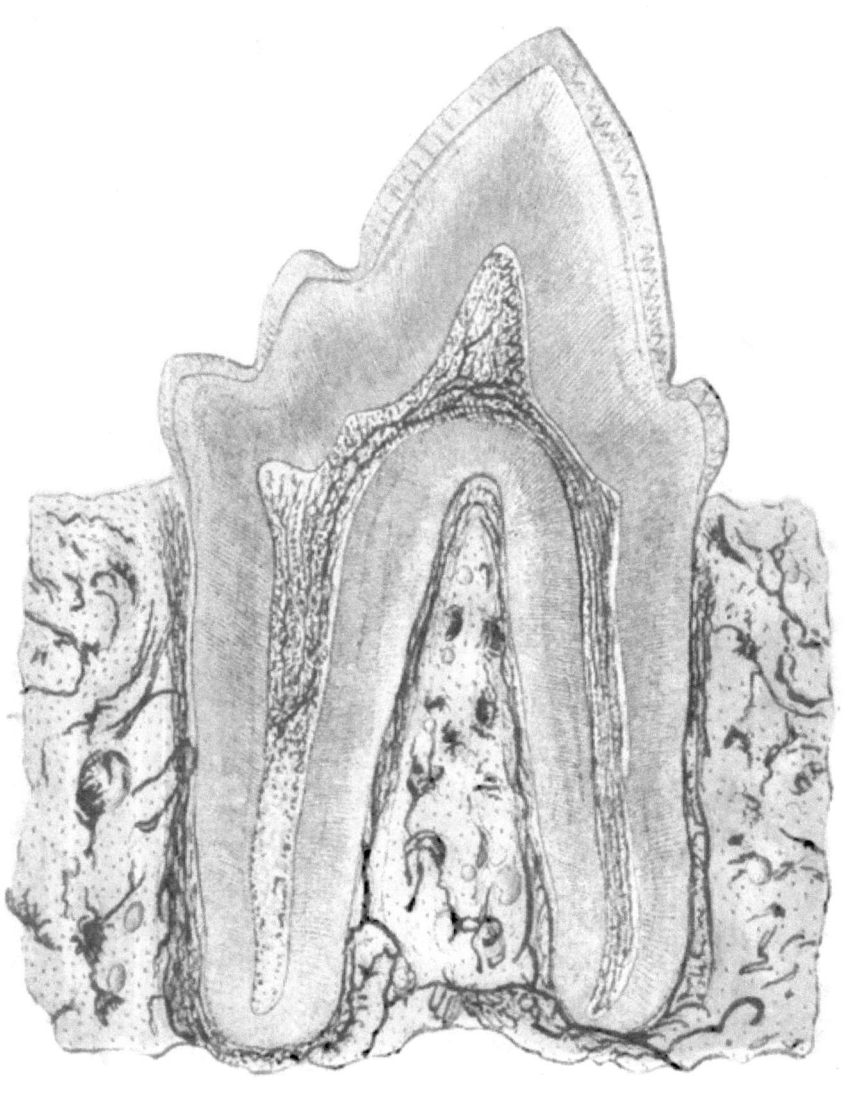

Section of molar tooth of a Cat (injected)

HISTOLOGICAL DEMONSTRATIONS:

A GUIDE TO THE MICROSCOPICAL EXAMINATION OF THE
ANIMAL TISSUES IN HEALTH AND DISEASE

FOR THE USE OF

THE MEDICAL AND VETERINARY PROFESSIONS.

BEING

THE SUBSTANCE OF LECTURES DELIVERED BY

GEORGE HARLEY, M.D., F.R.S.

PROFESSOR IN UNIVERSITY COLLEGE, LONDON, AND PHYSICIAN
TO UNIVERSITY COLLEGE HOSPITAL.

EDITED BY

GEORGE T. BROWN, M.R.C.V.S.

PROFESSOR OF VETERINARY MEDICINE, AND ONE OF THE INSPECTING OFFICERS
IN THE CATTLE PLAGUE DEPARTMENT OF THE PRIVY COUNCIL.

LONDON:
LONGMANS, GREEN, AND CO.
1866.

PREFACE.

IN OFFERING this work to the members of the Medical and Veterinary professions, with the hope of lessening the difficulties which obstruct the progress of the beginner in histological investigations, it is necessary to explain the circumstances under which it was written.

During an attendance on Dr. HARLEY's demonstrations in the physiological laboratory of University College, the observation of the facility with which objects were prepared for examination in the presence of the class, and the readiness with which the directions of the demonstrator were comprehended and carried into effect by the students, suggested to me the possibility of describing in an intelligible manner the method of instruction which was so successful in practice.

With Dr. HARLEY's concurrence and with the assistance of his original notes, which he placed at my disposal, the work was begun, and since then has gradually assumed proportions which were not at first contemplated, but which, with due regard to the completeness of the subject, could not be curtailed.

In attempting to convey the necessary instruction in the management of the microscope, I have thought it best to assume an absolute ignorance on the part of the reader, and to give the directions as nearly as possible in the same familiar manner in which Dr. HARLEY gives them to his class.

The plan of the work may be briefly explained.

In the first place the construction of the microscope is considered with reference to its optical and mechanical principles; and the various pieces of apparatus are shortly described.

Secondly, directions are given as to the arrangement of the instrument, the preparation of the slides, the position for the light, and the method of focussing the lenses.

Some general suggestions are then offered for the preparation of objects for examination; and lastly, under their several heads, the tissues of the animal body are concisely described, and the manner of preparing them for observation pointed out. In every instance a figure or diagram is given, with which the observer is enabled to compare the specimen, and thus ascertain how far his mode of preparation has been successful.

A large number of woodcuts have consequently been found necessary for the purpose of illustrating the appearance which every well prepared specimen should exhibit. Most of them were drawn from objects prepared exactly in the manner directed. Advantage has also been taken of Dr. Kölliker's kind permission to use some of the woodcuts from his large work on Microscopic Anatomy, and a few illustrations have been copied from Todd's *Cyclopædia of Anatomy*.

GEORGE T. BROWN.

LONDON: 1866.

CONTENTS.

	PAGE
HISTOLOGY	1
DESCRIPTION OF MICROSCOPE AND APPARATUS	1
Simple Microscope	1
Compound Microscope	2
Refraction	2
Reflection	2
Inflection or Diffraction	3
VARIOUS KINDS OF LENSES	3
SPHERICAL AND CHROMATIC ABERRATION	5
The Compound Microscope	7
Binocular Microscope	9
APPARATUS	11
MODE OF USING THE MICROSCOPE	16
Focussing	21
Preserving and Mounting Objects for the Microscope	22
Preservative Fluids	22
Colouring Fluids	22
Injected Preparations	25
ELEMENTARY TISSUES	27
Cells—their Nature and Function	27
Origin of Cells	29
Epithelial Cells	35
Tesselated Epithelium	35
Columnar Epithelium	36
Spheroidal Epithelium	37
Ciliated Epithelium	37
Lymph, Chyle, and Blood	38
Blood Crystals	47
Basement Membrane	49
Fibrous Tissue	49
Areolar Tissue, Cellular or Connective Tissue	53
Adipose Tissue	54

	PAGE
CARTILAGE	56
BONE	62
Development of Bone	67
MUSCLE	70
THE INTEGUMENT	78
Cutaneous Glands	84
HAIR	90
OF THE NAILS	97
HORN AND HOOF	100
TEETH	102
Dentine	104
Osteo-dentine	106
Enamel	106
Dental Pulp	107
DIGESTIVE CANAL	108
The Salivary Glands	111
The Mucous Membrane of the Stomach	112
Mucous Membrane of the Small Intestines	114
OF THE PANCREAS	120
THE LIVER	121
THE THYROID GLAND	123
THE THYMUS GLAND	124
THE SPLEEN	127
OF THE KIDNEYS	129
SUPRA-RENAL CAPSULES	133
LUNG	135
ARTERIES, VEINS, AND CAPILLARIES	138
Veins	140
Capillaries	140
OF THE NERVOUS SYSTEM	142
Gelatinous or Grey Fibres	145
Pacinian Bodies	148
The Brain and Spinal Cord	150
OF THE EYE	151
The Crystalline Lens	160
THE INTERNAL EAR	164

	PAGE
OF THE NOSE	165
ORGANS OF GENERATION	168
Milk	175
MORBID HISTOLOGY	177
MICROSCOPIC EXAMINATION OF MORBID FLUIDS	179
Pus	179
Serous Fluid	181
Milk	182
Urine	182
Urinary Deposits	182
Blood	184
CONCRETIONS	185
Biliary Calculi	185
Intestinal Concretions	186
Animal Concretions	187
Hair Concretions	187
Vegetable Concretions	188
Oat-hair Calculus	188
Starch Concretion	188
Urinary Concretions	188
DEGENERATION OF TEXTURES	189
Fatty Disease of Muscular Tissue	190
Fatty Degeneration of the Liver	190
Tuberculous Deposit	191
Typhous Matter	193
Pigmentary Degeneration	193
Bronzed Skin	194
ADVENTITIOUS PRODUCTS	196
Osseous Tumours	196
Exostosis	196
Osteophytes	198
Osteoma	199
Osteosarcoma	199
Enchondroma	199
Fibrous Tumours	200
Recurrent Fibroid	201
Indurated Chancre	202
Fatty Tumours (Lipomata)	202
Myeloid Growths	205
Myeloid Tumours	205
Hæmatoma	205

CONTENTS.

	PAGE
EPIDERMOID AND EPITHELIAL GROWTHS	206
Peculiar Forms of Disease affecting the Nails of the Human Subject	208
Elephantiasis	210
Epithelial Growths	211
Adenoid	212
Cystoid Tumours	213
Sebaceous Cysts	213
Glandular Cysts	213
Synovial Cysts	213
Compound Cysts	214
Colloid	215
Schirrus	216
Encephaloid. Cancer	218
Secondary Cancer	220
PARASITES	221
EPIZOA	222
Acari	223
Acarus folliculorum	223
Acarus scabiei	224
Ixoda	225
Pediculi	227
ENTOZOA	228
Cestoda—Tape-worms	229
Tænia solium	230
Tænia mediocanellata	332
Tænia marginata	233
Tænia echinococcus	234
Tænia serrata	235
Tænia crassicollis	235
Tænia cœnurus	236
Tænia elliptica—Tænia cucumerina	237
Bothriocephalus latus	237
Nematoda—Round Worms	239
Ascaris lumbricoides	239
Oxyuris vermicularis	241
Filaria bronchialis, or Strongylus bronchialis	241
Filaria oculi	243
Filaria Medinensis, or Guinea worm	243
Strongylus gigas	244
Strongylus paradoxus	244

ENTOZOA—*continued*.
    Trichocephalus dispar 244
    Trichina spiralis 245
    Sclerostoma syngamus 247
    Sclerostoma duodenale, or Anchylostomum duodenale . . 247
    Pentastoma 248
  Trematoda—Fluke-shaped Worms 248
    Distoma hepaticum, or Fasciola hepaticum 248
    Distoma lanceolatum 251
    Distoma ophthalmobium 251
    Distoma hæmatobium 251
    Distoma heterophyes 252

VEGETABLE PARASITES 252
  Fungi 253
    Oidium albicans 253
    Aspergillus 254
    Mycetoma Carteri 255
    Trichophyton tonsurans 256
    Microsporon andonini 257
    Microsporon mentagrophytes 257
    Microsporon furfur 258
    Achorion Schönleinii 258
  Algæ 259
    Cryptococcus cerevisiæ, Torula cerevisiæ, Yeast-plant . . 259
    Sarcina ventriculi 259
    Leptothrix buccalis 260

LIST

OF

ILLUSTRATIONS.

Section of molar tooth of a cat, injected	*Frontispiece*
Simple microscope: diagram	PAGE 1
Compound microscope: diagram	2
Action of double convex lens on parallel rays	3
Action of double convex lens on converging rays	4
Action of double convex lens on diverging rays	4
Action of plano-convex lens on parallel rays	4
Spherical aberration: diagram	5
Chromatic aberration: diagram	6
Section of the English achromatic combination	6
Compound microscope	7
Harley's binocular microscope	10
Box for apparatus	12
Manner of holding the glass slide	17
Method of applying the covering glass	19
Air-bubbles and oil-globules	20
Simple cells	27
Human cartilage cells	28
Ova of Ascaris nigrovenosa	29
Cartilage cells of full grown tadpole: semidiagram	30
Blood globules of chick, in act of division	30
Membrane from inner layers of an onion	31
Section of root of Iris Germanica	32
Starch granules from potato	32
Cells of black pigment of man	33
Cuticle of Iris, showing stomata	34
Epithelial cells from human oral cavity	36
Epithelium of intestinal villi of rabbit	36
Spheroidal epithelium from human bladder	37
Ciliated epithelium, from human trachea	38
Chyle from lacteals	39
Human blood globules	40
Human blood globules treated with water	42
Blood discs of the fowl	43
Blood discs of the frog	44
Blood discs of the fowl treated with tannin	45

LIST OF ILLUSTRATIONS.

	PAGE
Blood crystals	48
White fibrous tissue	50
Transverse section of tendon	51
Yellow elastic tissue from ligamentum nuchæ	52
Adipose tissue with fatty acid crystals	55
Cartilage of humerus of ox	57
Cartilage of rib of old man	58
Cartilage from mouse-ear	59
Yellow cartilage from horse's ear	60
White fibrous cartilage from patella of ox	60
Laminæ of decalcified bone	63
Transverse section of femur	64
Section from parietal bone	65
Longitudinal section from human femur	66
Granules of earthy matter from bone	67
Longitudinal section of ossifying cartilage from humerus of calf	68
Transverse section of ossifying cartilage	69
Fasciculi of muscle	70
Transverse section from human sterno-mastoid	71
Muscular fibres of horse	72
A muscular fibre breaking up into discs	73
Torn muscular fibre within untorn sarcolemma	73
Muscle of frog with nucleated cells	74
Primitive fibrils from primitive fasciculus	75
Vessels of muscular tissue	75
Non-striated elementary fibres from human colon	77
Vertical section through human skin	79
Perpendicular section through negro skin	80
Compound papillæ of the surface of the hand	81
Side-view and transverse section of papilla	82
Vessels of the papillæ	82
Perpendicular section through the scalp, with two hair-sacs	83
Sudoriferous glandular coil and its vessels	85
Perpendicular section of epidermis, showing sweat-duct	85
Sweat-ducts, highly magnified	86
Sebaceous glands from the nose	87
Glandular vesicle of sebacous gland	88
Section through skin of external auditory meatus	89
Hair and hair-follicle	90
Surface of hair, and epidermic plates	91
Transverse sections of human hair	92
Fibre-cells of the cortical substance of a hair	93
Hairs from various animals	94
Sections of hairs of various animals	95
Linen, cotton, woollen, and silk fibres	96
Transverse section of the nail and its bed	97
Transverse section of the nail, highly magnified	98

LIST OF ILLUSTRATIONS.

	PAGE
Longitudinal section of the nail and its bed	98
Nail-plates boiled with caustic soda	99
Transverse section of horse's hoof	101
Sections of human molar	102
Section of incisor of horse	103
Dentine and cement from human incisor	104
Transverse section of human dentinal canals	105
Section of tooth of saw-fish	105
Human dentine and enamel	106
Extremities of enamel-fibres of calf	106
Cement and dentine of old human tooth	107
Diagram of two ducts of a lobule of a mucous gland	108
Follicular gland from root of human tongue	108
Upper surface of human tongue	109
Section of human papilla circumvallata	110
Fungiform papilla	110
Filiform papilla	111
Œsophageal glands of man	112
Perpendicular section through coats of pig's stomach	113
Pyloric and cardiac tubes from dog's stomach	114
Section of jejunum	115
Vessels of villi of mouse	115
Villi of calf	116
Racemose gland in section of duodenum	117
Agminate follicles	118
Portion of a cluster of agminate follicles	118
Plan and section of an agminate follicle	119
Lobule or acinus of mouse's pancreas	120
Interlobular spaces, fissures, and veins of liver	121
Hepatic cells of man	122
Termination of interlobular duct in pig's liver	123
Gland-vesicles from thyroid gland of a child	124
Piece of the thymus of a calf	125
Transverse section through thymus of a child	126
Transverse section through spleen of ox	127
Various cells from the spleen	128
Malpighian corpuscle from the spleen	129
Section of child's kidney	130
Urinary tubules of man	131
Human Malpighian corpuscle with urinary tubule	132
Transverse section through tubules of kidney	133
Transverse section of human suprarenal capsule	134
Human pulmonary vesicle	136
Pulmonary lobules: semidiagram	137
Capillary network of human pulmonary vesicles	138
Longitudinal section of aorta of horse	139
Finest vessels of the arterial side, from human brain	141

LIST OF ILLUSTRATIONS.

	PAGE
Branch of nerve from the frog	143
Nerve fibres	144
Grey nerve fibres	146
Nerve vesicles from human Gasserian ganglion	147
Caudate nerve vesicles from human brain	148
Human Pacinian body	149
Transverse section of human spinal cord	151
Section of cornea and sclerotic	152
Nerves of cornea of rabbit	154
Cells from the stroma of human choroid	155
Vessels of choroid and iris of a child	156
Nerves of iris of white rabbit	157
Perpendicular section of human retina	158
Elements of bacillar layer, from human retina	159
Nerve cells from human retina	160
Lenticular fibres of ox and man	161
Simplest form of crystalline lens	162
Lens with transverse septum	162
Lens with three septa	163
Central point of lens of salamander	163
Transverse section through spiral lamella of cochlea	164
Membrane of the nose	166
Olfactory tubes of the ox	167
Seminal tubule of man	168
Development of spermatic filaments of bull	169
Human spermatozoa	170
Spermatozoa of Nepa cinerea	170
Spermatozoa of perch	171
Spermatozoa of cock	171
Mother-cell with spermatozoa from Fringilla	171
Spermatozoa of rat and mouse	172
Human ovulum	173
Sections of corpora lutea	174
Lobules of lacteal gland	175
Morphological elements of milk	176
Pus and mucus corpuscles	181
Crystals of uric acid	182
Oxalate of lime from urine	183
Crystals of triple phosphate	183
Urinary deposits in disease	184
Fractured surface of biliary calculus	186
Section of triple phosphate calculus	187
Oat-hair calculus	188
Fatty degeneration of the heart	190
Fatty degeneration of the liver	191
Tubercle	192
Fibres of lens in black cataract	194

LIST OF ILLUSTRATIONS.

	PAGE
Section of bronzed skin	195
Sections of diseased bone in horse	197
Foliaceous osteophyte	199
Section of enchondroma	200
Fibrous tumour	202
Section of reticulated fatty tumour	204
Section of erectile tumour	206
Scrapings from thrush, foot-rot, and canker	207
Disease of nail and horn	209
Sections of integument in elephantiasis	210
Elements of epithelial cancer	211
Adenoid tumour from mammary gland	212
Proliferous ovarian cyst	214
Colloid cancer of the ovary	215
Schirrous growth from the mammary gland	216
Elements of encephaloid tumour	219
Acarus folliculorum	223
Acarus ovis and A. scabiei	225
Sheep and dog ticks	226
Head, body, and crab lice from man	227
Tænia solium	230
Tænia mediocanellata	232
Echinococcus hominis	234
Tænia crassicollis and its larva, Cysticercus fasciolaris	236
Bothriocephalus latus	238
Ascaris megalocephalus of the horse	240
Oxyuris vermicularis	241
Strongylus filaria	242
Young of Guinea worm	243
Trichocephalus dispar	245
Cysts of Trichina spiralis in situ	246
Separate magnified cyst of Trichina	246
Mature Trichina	247
Liver fluke	249
Development and generations of Distoma	250
Distoma hæmatobium and ova	252
Oidium albicans from the mouth	254
Aspergillus from the ear	255
Foot-fungus of India	255
Hair with spores of Trichophyton tonsurans	256
Microsporon mentagrophytes	257
Microsporon furfur	258
Achorion Schönleinii	258
Torula cerevisiæ	259
Various forms of Sarcina ventriculi	260
Algæ from the mouth and larynx	261

INTRODUCTION.

No ARGUMENTS are needed in the present day to prove the vast importance of microscopical research. Not only is all natural science indebted to the microscope; but art and commerce are compelled to have recourse to its assistance to solve problems that could hardly receive a solution without it:—who, for example, would a few years ago have dared even to imagine that the microscope would enable the artisan to distinguish Bessemer's and other kinds of iron more exactly than by any other means? Notwithstanding the value of the instrument, however, it is not without opponents, who are ready enough to charge upon it all the errors that have from time to time been published by enthusiastic but injudicious friends, whose zeal has probably far exceeded their powers of observation, and whose anxiety to give the world the benefit of their fancied discoveries, has done more damage to the cause they sought to defend than all the antagonism of those who see in the instrument only a toy to amuse an idle hour.

The worker with the microscope does not pretend to unravel enigmas as if by a magic crystal; he certainly trusts to it to extend his powers of observation when his unaided eyesight fails to convey further information. Up to this point however he uses all the ordinary means of examination before availing himself of a magnifying power, and even then, instead of endeavouring to enlarge the image of the object several million times, in accordance with a favourite popular delusion, he never uses a high power until he finds a low one insufficient to give him the information required. A low power will often indeed teach more than a high one.

Many important improvements have been made of late years both in the mechanical and optical parts of the instrument. Every possible movement can now be given to the object by means of levers, screws, or rack and pinion; while in the matter of objectives, there would seem to have been a contest between the optician and the observer, in which the former has for the present gained a very decided victory by the production of a glass of $\frac{1}{50}$ of an inch focus.

Some years since Mr. Ross succeeded in producing an excellent $\frac{1}{12}$. Messrs. Powell and Lealand followed quickly with a $\frac{1}{16}$. Messrs. Smith, Beck, and Beck have the credit of manufacturing a $\frac{1}{16}$, while the crowning point seemed to have been reached when Messrs. Powell and Lealand gave to the scientific world a $\frac{1}{25}$ so admirably constructed as to be available for the examination of ordinary objects with a covering glass of ·005 of an inch in thickness. These makers, to whom too much praise can hardly be accorded, have recently succeeded in constructing a very effective $\frac{1}{50}$ of an inch objective.

With such perfect appliances, on the one hand, and moderate-priced instruments on the other (even the latter being far better than those with which in former times the most important discoveries were made), there is not only enough to stimulate a love of research, but also to justify a hope that the microscope will ere long be reckoned among the necessary implements of every scientific man, and its use an indispensable part of the education of the student.

Referring more immediately to our own special subject, it is difficult to understand how the anatomist and pathologist can pursue their inquiries without the aid of the microscope.

In its absence the demonstration of the elements of the simplest structure is impossible, and the nature of a morbid deposit can be only a matter of conjecture, while with the assistance of the instrument there is reason to hope that many obscure pathological problems, the number of which is constantly being increased by new accessions, might be elucidated; and were the employment of the microscope to become universal among the professors of medical science, important additions to our knowledge might fairly be anticipated.

HISTOLOGY.

DESCRIPTION OF MICROSCOPE AND APPARATUS.

In the limited space available for the discussion of preliminaries, only a very brief explanation of the microscope and apparatus is possible; nor, indeed, would the most complete dissertation amount to more than a reproduction of the information contained in the larger works upon the subject.

A general knowledge of the instrument is necessary for the student, and it will be our object to convey the requisite information in the simplest and most familiar manner.

As the word implies (μικρός, *small*, and σκοπέω, *to view*), every glass or lens which presents to the eye a magnified image of an object constitutes a microscope.

SIMPLE MICROSCOPE.—The most simple form of microscope is a single lens, or combination of several lenses so arranged that the magnified image of the object is directly presented to the eye; all such instruments, whatever the manner of their construction, are simple microscopes. The illustration represents the action of this kind of microscope with one lens.

Fig. 1.

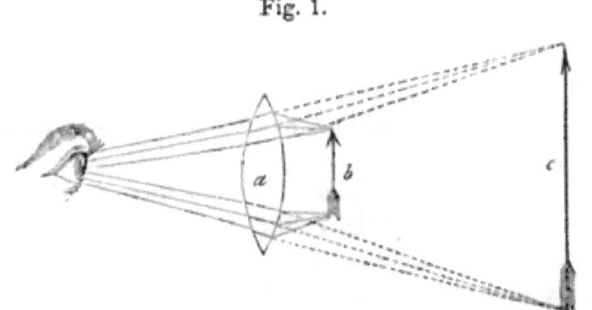

Simple Microscope. *a.* Lens. *b.* Actual object. *c.* Apparent position and size of object to the eye of the observer.

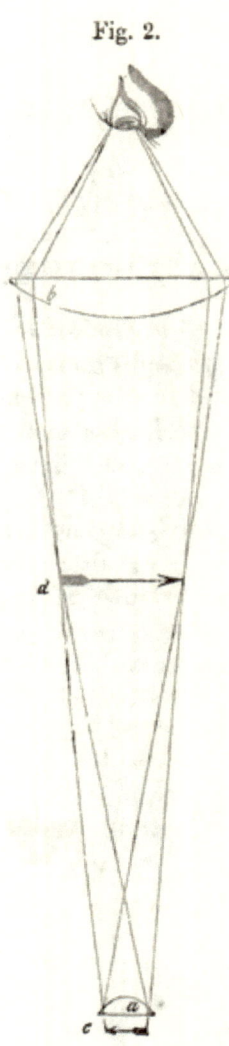

Fig. 2.

COMPOUND MICROSCOPE.

A COMPOUND MICROSCOPE, on the other hand requires, that the lenses, fig. 2, *a, b,* whatever their number, shall be so arranged that the magnified image, *d,* of the object, *c,* presented by one set of glasses shall be, as it were, intercepted and again magnified by another set, before reaching the eye of the observer, as shown in the diagram.

The magnified image, *d,* is always inverted in consequence of the rays of light which proceed from any part of an object below its centre coming to a focus above it, and *vice versâ.*

The magnifying power of a substance depends upon the amount of refraction which the rays of light undergo in passing through it.

REFRACTION.—Rays of light passing from a rarer to a denser medium at an angle are refracted *towards* a line drawn perpendicularly to a plane which divides them.

Rays of light passing from a denser to a rarer medium at an angle are refracted *from* the perpendicular.

Perpendicular rays always pass unchanged.

The lines of the angles of incidence and refraction, viz., the angles which the rays form with the perpendicular line before and after refraction, bear to each other a certain ratio for each substance, which is known as the index of refraction.

REFLECTION.—Rays of light falling upon an opaque polished plane surface at an angle are reflected at the same angle; thus the angle of incidence and the angle of reflection are equal.

If the surface is concave the reflected rays converge, if convex the rays diverge.

INFLECTION OR DIFFRACTION.—Rays of light passing near the edges of bodies produce the appearance of a shaded and sometimes coloured fringe or band, instead of a definite outline. This *diffraction band* may be mistaken for an actual portion of the object under examination, particularly in very delicate investigations with high powers, and it is only by experience that such errors can be avoided.

VARIOUS KINDS OF LENSES.

LENSES used in the construction of microscopes are chiefly convex, the concave being used only to modify the course of the rays passing through the convex.

Rays of light passing through convex lenses are converged and brought to a focus at a certain point depending upon the refractive power of the material of which the lens is formed, and its degree of curvature.

The refractive power of common glass is 1·5.

Double convex lenses cause parallel rays passing through them to converge to a focus at the centre of the circle of which the lens is a segment; and conversely, rays diverging from the centre are rendered parallel.

Fig. 3.

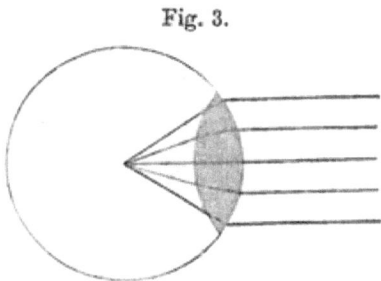

Parallel rays falling on a double convex lens brought to a focus in its centre; rays diverging from such a point rendered parallel.

Converging rays passing through a double convex lens are brought to a focus at a point in front of the centre; and conversely, rays diverging from that point will still diverge, though in a diminished degree.

HISTOLOGY.

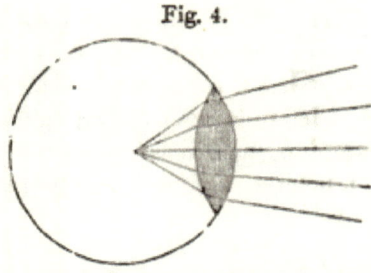

Fig. 4.

Rays already converging brought to a focus nearer than the centre; and rays diverging from such a point, still diverging in a diminished degree.

Rays diverging from a point behind the principal focus on either side are brought to a focus beyond it. If the point of divergence be within the circle of curvature the focus of convergence is beyond it, and *vice versâ*.

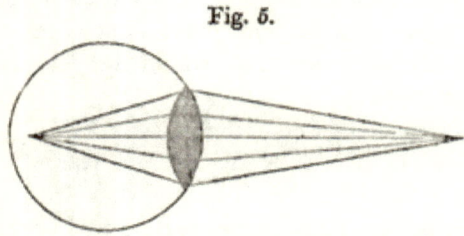

Fig. 5.

Rays diverging from points more distant than the principal focus on either side brought to a focus beyond it.

Plano-convex lenses cause the rays of light passing through them to be brought to a focus at the distance of the diameter; and conversely, rays diverging from that point are rendered parallel.

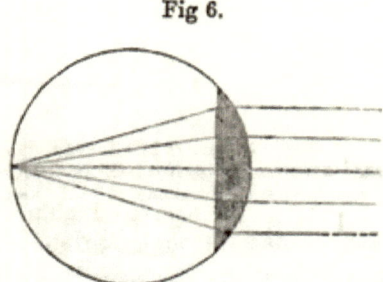

Fig 6.

Parallel rays falling on a plano-convex lens brought to a focus at the distance of its diameter, and vice versâ.

Concave lenses cause parallel rays to diverge as from a principal focus, which in these lenses is called a *negative focus*. For a plano-convex lens the negative focus will be at the distance of the diameter of the sphere; and for a double concave lens the centre of that sphere.

SPHERICAL AND CHROMATIC ABERRATION.

On looking into some microscopes the edges of the object appear less distinct than its centre, or it may be surrounded by a rainbow-coloured ring.

The first of these defects is due to spherical aberration; the second to chromatic aberration.

SPHERICAL ABERRATION is caused by the central rays of light not arriving at a focus so soon as the peripheral rays.

Fig 7.

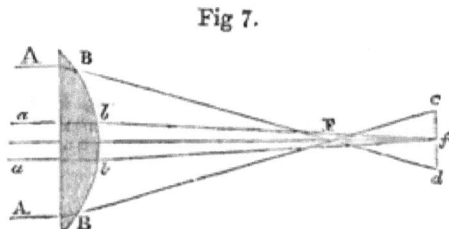

SPHERICAL ABERRATION.

A B, rays falling on the periphery of the lens; F, focus of these; a, b, rays falling nearer the centre; f, more distant focus of these.

The remedy for this defect is to increase the curve of the central part of the lens, or to cut off the peripheral rays by the intervention of a diaphragm.

CHROMATIC ABERRATION depends not so much upon the form, as upon the material of which the lenses are made. Some substances cause white light to be separated into its component rays as in the common prism; and, as the different colours have different focal points, it will be found that according as the lens is focussed, the violet, blue, green, red, or yellow will prevail.

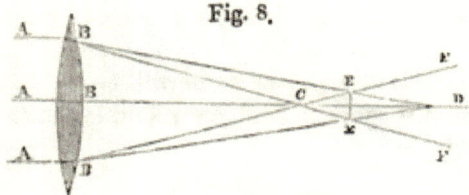

Fig. 8.

CHROMATIC ABERRATION.

A B, rays of white light refracted by a convex lens; C, the focus of the violet rays, which then cross and diverge towards E F; D, the focus of the red rays which are crossed at the points E E, by the violet; the middle point of this line is the mean focus, or focus of least aberration.

To remedy this very serious defect two materials are employed, having different dispersive powers, so that one may neutralise the error produced by the other.

One substance having too much, and the other too little dispersive power, the plus of the one, and the minus of the other, together give nil.

In constructing the better class of objectives a double convex lens of crown glass is neatly fitted into a plano-concave of flint glass; and to their surfaces is given such a degree of curvature that the excess of aberration in the one is compensated by the minus of the other.

Three sets of lenses so constructed are generally combined in a good objective, fig. 9, P, M, A.

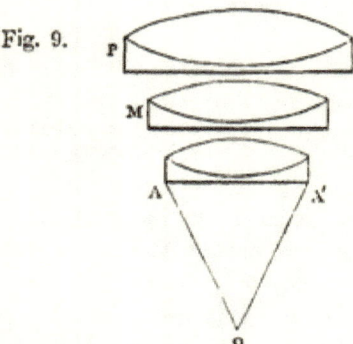

Fig. 9.

SECTION OF THE ENGLISH ACHROMATIC COMBINATION.

Aberration is still further corrected in the eye-piece, which is composed of two lenses, one called the 'eye glass,' from being next the eye of the observer, and the other termed the 'field glass,' from being nearer the field.

The field glass is very important, as it not only tends to correct aberration, but also causes the rays of light from the object glass to converge, so that none, or very few of them, are lost; thus rendering the image more clear and distinct than it would otherwise be.

The Compound Microscope.

The mechanical part of the instrument will be understood by reference to the accompanying drawing.

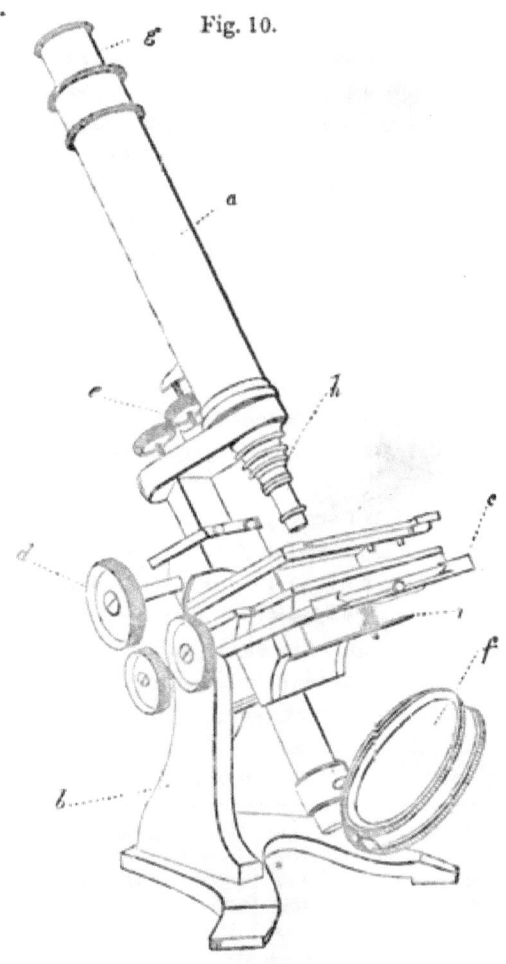

Fig. 10.

COMPOUND MICROSCOPE.

The lenses (g h) are fixed in a tube (a) set in a solid frame (b), to which is attached a stage (c) on which the object may be securely placed. Arrangements are made by means of rack and pinion for bringing the tube near to, or removing it from the object plate or stage, both by the coarse and fine adjustments (d e).

For the illumination of transparent objects a mirror (f) is placed under the stage upon a movable arm, which allows it to be readily adjusted.

Every part of the instrument should be firm, and vibrate equally; and all the movements should be smooth and free from catches or jerks.

All the necessary mechanism of the microscope is extremely simple, and easily and quickly used. The luxurious complications of the most expensive instruments are little valued by the worker in the laboratory to whom time is important.

Even the recent improvements in the object-glasses of high power, which are made to transmit large pencils of light through their large angular apertures, and thus enable the observer to distinguish very delicate objects, such as the lines upon Diatomaceæ, are not adapted for ordinary scientific research; certainly not for every-day physiological or pathological investigations; not only because their adjustment is, of necessity, extremely delicate, but for the further reason that their power of penetration is impaired by the arrangement which improves their power of definition.

The object-glasses are designated according to the focal distance of a single lens of the same magnifying power: thus a two-inch objective is understood to be a combination which has the magnifying power of a single lens whose focal point is two inches from the object, and so in reference to the other powers. The following table will convey an idea of the magnifying power of each object-glass with the different eye-pieces.

EYE-PIECES	OBJECT-GLASSES								
	2 in.	1 in.	$\frac{1}{2}$ in.	$\frac{1}{4}$ in.	$\frac{1}{8}$ in.	$\frac{1}{12}$ in.	$\frac{1}{16}$ in.	$\frac{1}{25}$ in.	$\frac{1}{50}$ in.
A	20	60	100	220	420	600	800	1250	2500
B	30	80	130	350	670	870	1184	1850	3700
C	40	100	180	500	900	1400	1600	2500	5000

These measurements are linear, but, with a view to astonish the public, it is common to speak of the superficial enlargement, which is the square of the linear; thus, if the quarter-inch magnifies an object about 200 diameters, or 200 linears, its superficial measurement will be obviously the square of the 200 linear, viz. 40,000.

The magnifying power of the eye-piece used must be added to the magnifying power of the objective in estimating the enlargement of the image of an object.

By the aid of different eye-pieces an extensive range of magnifying power may be obtained; for example, the two-inch objective with a deep eye-piece will give the same amplification as the quarter objective with the ordinary eye-piece; and for certain observations, the combination of low objectives with deep eye-pieces is by some considered to be advantageous.

Binocular Microscope.

The employment of the additional tube and eye-piece with Wenham's prism is now so general that little need be said in its favour.

The peculiar stereoscopic effects produced, particularly with the low powers, give a natural appearance to objects, not attainable with the single microscope. At the same time the binocular arrangement in no way interferes with the use of the instrument as a single microscope.

The accompanying drawing, fig. 11, represents the microscope recommended by Dr. Harley to his students. It is constructed on his own plan, and possesses the advantage of having two object glasses and various pieces of apparatus, so arranged as to be brought into use without a moment's loss of time. It is specially adapted to the requirements of the busy man.

The microscope as is here seen is fixed into the bottom of the mahogany box which forms at the same time the stand. Round it a groove is run to receive the lip of a glass shade. The eye-pieces are supplied with shades (a, a) to protect the eyes.

These are a great comfort to the observer when he is using the instrument for any length of time.

At the end of the transverse arm (f) is the box which contains

HISTOLOGY.

Fig. 11.

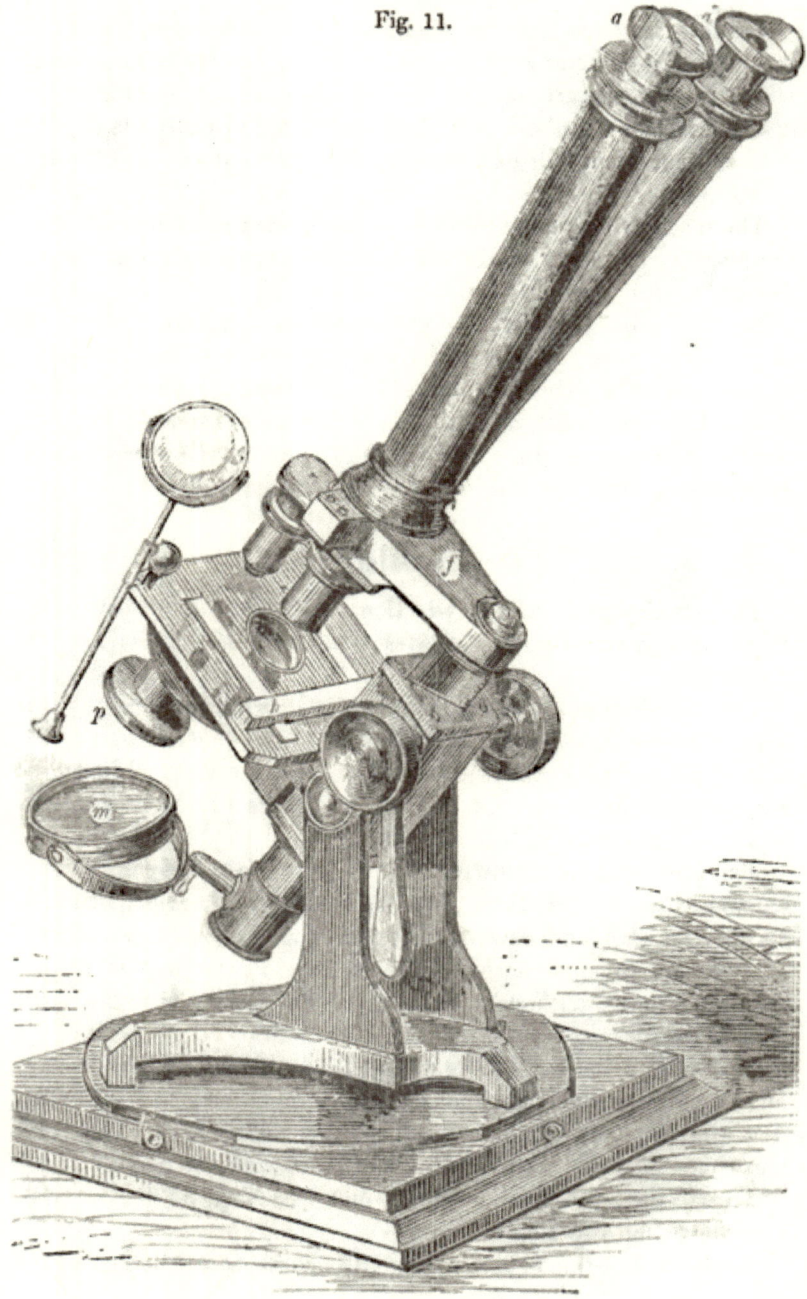

HARLEY'S BINOCULAR MICROSCOPE.

both Wenham's binocular prism, and the analyser of the polariscope; and by merely drawing it a little out, or pushing it further in, the instrument can be instantly changed from a binocular to a monocular, and still further to a polarising microscope.

Immediately beneath f are the two objectives, a quarter, and an inch; so that in order to change the power, all that is necessary is to slide them backwards or forwards. Moreover, these are fitted with the universal screw, so that either of them may be detached, as in an ordinary instrument, and a one-eighth, a one-twenty-fifth, or any other power, put in its place at the option of the observer. The instrument is fitted with a coarse and fine adjustment, and has the additional advantage of a magnetic stage, in the crossbar (h) of which is a groove, in order that the observer may apply a Maltwood's finder, as in large instruments possessing movable stages. Beneath the stage is seen the polariser (p) fitted into the circular diaphragm.

The double mirror (m) possesses a triple joint, so that it can be applied obliquely in all directions.

APPARATUS.

BESIDES the microscope with high and low powers some simple apparatus will be required for the preparation of the objects for examination, for example: A scalpel or two; a good razor kept in perfect order, for making sections; straight and curved forceps; a pair of curved scissors; some needles fixed in wooden handles; two or three glass rods and pieces of glass tube; and some glass slides (3 in. by 1 in.), with some pieces of thin covering glass half an inch square.

A very convenient box, such as is employed in the University College Laboratory, is represented in fig. 12, having divisions to hold a tumbler of water (A) for washing the glass slides, a bottle of distilled water (B), bottles of solution of potass (C), and dilute acetic acid (D); and also compartments for the needles, rods, and dipping tubes (E), as well as for the slides (F) and covering glasses (G).

Fig. 12.

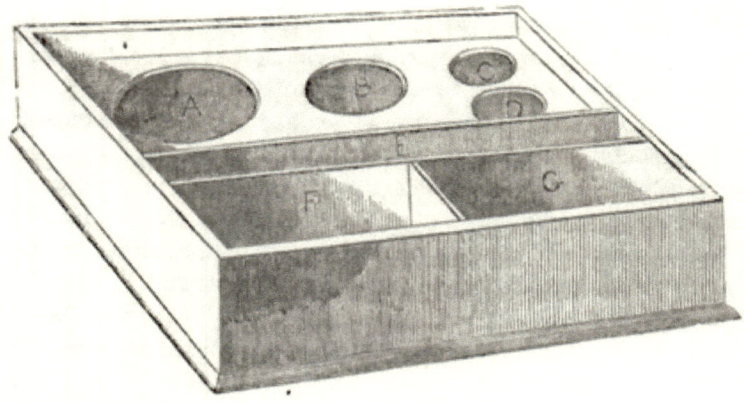

ADDITIONAL APPARATUS.—*Bull's-eye condenser* is employed for the illumination of opaque objects, and may be either fitted to the stage or fixed to a separate stand. It consists of a plano-convex lens, which is placed between the object and the lamp, and serves to concentrate the rays of light upon the upper surface of the object.

Side Reflector consists of a concave silver speculum attached by an arm to the stage of the microscope. It is used in conjunction with the bull's-eye condenser, by which the rays of light are thrown on to the reflector, and by it directed to the object in an oblique direction, often rendering minute markings apparent, which would otherwise not be distinguishable.

Lieberkühn is a small silvered speculum or reflector, attached to the object-glass at such a distance from the object that the light thrown upon it by the mirror is brought to a focus at the exact point of focus of the objective.

Dark wells or stops are generally employed with the Lieberkühn to hold the object, and also to prevent the passage of any rays of light to it save those reflected from the Lieberkühn.

Achromatic condenser.—This instrument consists of an achromatic combination of lenses, for the purpose of condensing light upon transparent objects, and at the same time, by means of a movable diaphragm, with apertures of various shapes and sizes, to cut off all superfluous rays, whether central or peripheral.

The apparatus is usually adapted to a secondary stage, as it requires accurate adjustment.

Camera Lucida is employed for sketching an object where great accuracy of outline is required. Three forms of apparatus are in use for the purpose.

1. *Wollaston's prism*, by means of which rays of light proceeding from the object are bent at right angles, and enter the eye as if they came from the paper placed underneath the instrument. In using this or any other form of camera, the microscope must be placed in a horizontal position, the object focussed, and the cap of the eye-piece removed, to allow the camera to be put on in its stead. A sheet of paper should then be placed on the table under the camera, and the eye brought over the edge of the prism and directed downwards. The object will now appear on the paper, and may be accurately traced by means of a pencil with a blackened point, which is more readily seen than a light coloured one.

2. *Sœmmering's speculum* is a polished steel disc set at an angle of 45 degrees, and fitted to the eye-piece by a brass clip. Is employed in the same way, and is by some preferred to the prism.

3. *Neutral glass Reflector.*—A piece of neutral glass fixed at an angle of 45 degrees, and attached to the eye-piece by a brass clip, answers the purpose very perfectly, and is cheaper than the other forms. It must be borne in mind that the image is enlarged in proportion to the distance of the camera from the sketching paper, consequently if very large drawings are required the microscope may be adjusted so that the end of the tube shall extend beyond the edge of the table, while the paper is placed on the ground. A pencil tied on to a stick of sufficient length will then answer very well for sketching the object.

The Micrometer, for measuring the size of an object, consists of a piece of glass graduated to $\frac{1}{100}$ and $\frac{1}{1000}$ of an inch, and fixed in a metal or cardboard slide, like the ordinary glass slide. This form is termed the stage micrometer.

Another method consists in graduating one of the glasses of the eye-piece, or inserting a graduated slide between the eye-glass and the field-glass.

In using the stage micrometer the lines are first to be brought into focus; the camera is then to be applied to the eye-piece, and the microscope arranged so that the divisions on the micrometer can be sketched. The object must next be substituted for the micrometer, and the image compared with the divisions previously drawn on the paper.

The eye-piece micrometer, before being employed to determine the measurement of an object, requires that the value of the divisions, with one or more of the objectives, should be found by comparison with the stage micrometer. If, for example, any four divisions of the eye-piece micrometer correspond with the $\frac{1}{100}$ part of an inch on the stage micrometer, each of those four divisions is of the value of $\frac{1}{400}$ of an inch. The correction is readily made when the microscope possesses a graduated draw-tube.

The Parabolic Reflector is an apparatus for illuminating by oblique light. It consists of a piece of glass worked to the form of a parabola, having a centre aperture, in which a movable stop is adjusted. By its employment very delicate markings can be distinguished, brilliantly illuminated upon a dark field.

The parabolic reflector may be adapted to the under surface of the stage, or fitted to the substage, as it usually is in the larger instruments.

In using the parabola the flat side of the mirror should be employed; and with artificial light it is necessary to place the condenser between the lamp and the mirror, to render the rays of light as nearly parallel as possible. For this purpose the condensing lens must be placed very near the lamp with its flat side to the light, so as to throw parallel rays upon the mirror, which transmits them through the parabola, from which they are reflected in such a way that no rays can enter the object-glass except those that are intercepted by the object, consequently the object only is illuminated and the field remains dark.

The *Polariscope* is composed of two parts: a polarising medium such as a prism of Iceland spar, which is placed under the stage, between the mirror and the object, and an analysing medium which is screwed into the tube between the object-glass and the eye. The employment of these darkens and contracts the field considerably; but they are, nevertheless, of great service, especially in the examination of crystalline substances, the true nature of which is sometimes discernible from their polariscopic characters alone. In using the polariscope, the object is to be focussed in the ordinary manner, and that portion of the instrument which is placed under the stage is made to revolve on its axis, by means of the finger and thumb, and the play of colours it produces noted.

The Erector is a combination of lenses arranged to screw into the draw-tube of the microscope to correct the inversion of the image, and cause it to appear in the natural position; it is principally useful for dissecting with the compound microscope; but it also affords a considerable range of magnifying power when the draw-tube is used to increase the distance between the object and the eye.

The Microscope Lamp.—A good steady light is absolutely indispensable to the success of microscopic observation. The lamp made and sold by Mr. Baker, 244, Holborn, combines efficiency with economy; its price is 10s., and the paraffin which it burns is comparatively inexpensive.

Valentin's Knife is an instrument for making sections of soft substances; it is constructed of two scalpel-like blades which can be adjusted at different distances from each other by means of a screw, according to the thickness of the section required. The tissue should be fixed under water, and the section when made floated on to the glass slide without being removed from the fluid.

The *Compressorium* may be used for securing live objects and preventing or limiting their movements, or for the purpose of compressing substances that are too thick to be examined as transparent objects without such preparation.

This apparatus is made in various forms, but it essentially consists of a brass plate into which a piece of circular glass is fitted to receive the object, and a brass ring containing the thin covering glass; the two parts are so adjusted together that the cover can be made to press upon the glass on which the object is placed, by the action of one or more screws.

Animalcule Cage is useful for examining animalcules in water, or for fixing small animals that would otherwise escape from the field of the microscope. The instrument is composed of a piece of brass plate rather larger than an ordinary slide, in the centre of which is a ring having a circular piece of glass at the upper part, surrounded by a groove to receive the excess of fluid. The covering glass is fitted into a brass cap which is made to slide or screw over the part on which the object is placed.

Zoophyte trough is a glass trough very convenient for the examination of large objects, plants, or animals in water.

The space in which the object is confined may be regulated by a whalebone spring and a counteracting wedge of wood or ivory.

In using the glass trough the microscope must be placed in a position nearly horizontal.

Frog Plate for examining the circulation in the web of the frog's foot, consists of a wooden or brass plate large enough to accommodate the body of the animal enclosed in a bag which is to be secured to the broad end of the plate by means of tape. The small end is provided with an aperture fitted with a piece of glass over which the web is to be stretched.

Numerous pins and holes are arranged at the sides of the plate for the purpose of making the necessary fastenings.

Stage Forceps are sometimes fitted to one side of the stage for the purpose of holding small opaque objects, living or dead.

The instrument is always arranged to move in any direction, so that the object may be viewed in every position when illuminated by the bull's-eye condenser or the mirror beneath the stage.

Many other kinds of apparatus might be added to the list, and new forms are constantly being manufactured to meet the requirements, real or fancied, of a large class of observers. The student, however, is advised to master the art of working with simple apparatus, and to add the more complicated as he finds them to be necessary.

MODE OF USING THE MICROSCOPE.

THE student being seated at a low table, with his microscope conveniently placed towards his left hand, should first arrange his lamp in front and a little to the left of the mirror, and then incline the microscope tube until he can look into it comfortably with his left eye; taking particular care to keep the other eye open. The habit of closing one eye in using the monocular microscope is decidedly objectionable, and has no one advantage save that at first it is easier to yield to the tendency than to resist it, but a very little perseverance will suffice to overcome the difficulty.

The next step is to illuminate the field; this can be most readily accomplished by first lowering the object-glass to within a quarter of an inch of the stage and then adjusting the mirror until the light can be seen reflected upon the end of the object-glass. On looking into the microscope, the field will be seen to be perfectly illuminated, and by a little further adjustment of the mirror any required degree of light may be obtained.

The strongest light is reflected from the concave side of the mirror, that from the flat side is more diffused and less intense.

Oblique light can be obtained by turning the mirror on one side, and then adjusting it so as to illuminate the field from that position.

The next point is to clean the glass slide; for this purpose it must be taken by its edges between the fore-finger and thumb of the left hand; thus avoiding to touch the under or upper surface,

Fig 13.

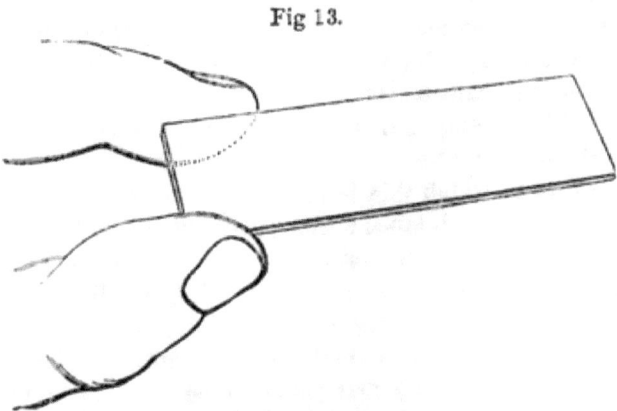

MANNER OF HOLDING THE SLIDE.

which are certain to be soiled by the fingers, however clean the hand may be. The slide, held by its edges, must be dipped into the tumbler of water standing conveniently by, and then, still being held by the edges, drawn a few times through a soft cotton cloth, a piece of good cotton sheeting for example, which ought to be hanging from the table within easy reach. The slide is thus rapidly cleaned and polished, and may be rested against the stand of the microscope by one end, while the covering glass is prepared. This must likewise be held by its edges, dipped into the water, and dried by being rubbed in the cloth between the finger and thumb; in this way a breakage even of this delicate glass is hardly possible.

The covering glass, being ready for use, ought to be placed on some easily remembered spot, previously determined upon, and habitually used for the purpose, otherwise much time will be often lost in the search for so small a piece of glass, whose situation is doubtful. Any part of the stand of the microscope will do, providing the locality be invariably the same, and always easily recognised.

The value of all these apparently trivial directions will be best appreciated by him who neglects them for a while. Saving of time is the great object to be kept in view; and when the mind is concentrated upon the elucidation of some obscure point, nothing is more annoying than the necessity for thinking of the whereabouts of the apparatus; whereas, by perseverance in a definite method of manipulation, the necessary movements are performed mechanically, and the attention is left free to be fixed upon more important matters.

Every thing being now ready, the preparation of the object will be the next step.

The mode in which this is to be accomplished will vary with the nature of the substance; the special methods, however, will be indicated when treating of the separate tissues; at present we have only to lay down some general rules, which may be considered applicable to the majority of cases.

Suppose, for example, it is intended to examine a piece of simple fibrous tissue. The first thing to be done is to place, by means of a glass rod, a drop of distilled water in the centre of the clean slide; next cut from the mass, by means of scissors, the smallest possible piece of fibrous tissue, and let it fall into the centre of the drop of water on the slide. Now place the slide on a dark ground, in order that the fragment of white tissue may be made more distinct by contrast; take one of the needles in the left hand, and fix it into the object, resting the arms comfortably but firmly on the table.

Having thus anchored the object, commence to teaze out its edges with the point of another needle held in the right hand, and gradually go on tearing the fragment into shreds until scarcely a visible particle of it remains; in fact, it is to be teazed out exactly on the same principle as the housewife teazes out a piece of calico, thread by thread, fibre by fibre. This feat accomplished, which, simple as it appears to be, requires some little perseverance to do well, lay aside the needles and pick up the covering glass, holding it by its edges between the finger and thumb, breathe on what you intend to make its under surface, in order to cover it with a film of watery vapour; then rest one of the edges upon the slide, close to the drop of water, and slowly depress the other end gently, as if you were gradually shutting down the lid of a box, until the drop of water is covered over, and the preparation of fibrous tissue with it.

Fig. 14.

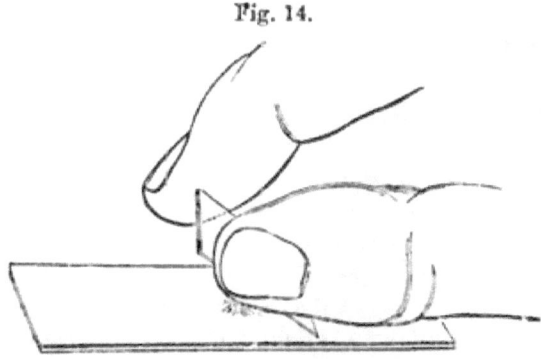

METHOD OF APPLYING THE COVERING GLASS.

The covering glass is thus lowered in order that the water may rise on the glass by capillary attraction, and drive the air before it.

If the covering glass is always applied carefully, according to the preceding directions, the observer will never be troubled with air-bubbles, which are the bugbears of all beginners.

In order that the student may recognise these obtrusive objects at once, and be enabled to compare them with oil-globules, which also commonly occur as accidental objects in specimens of animal tissues, the illustration represents air-bubbles and oil-globules in juxtaposition.

Fig. 15.

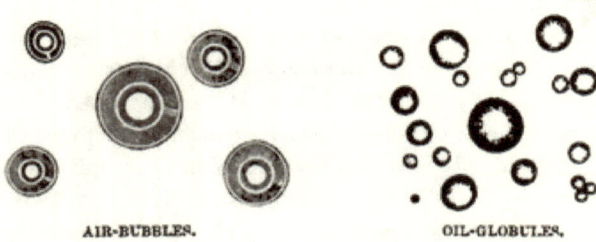

AIR-BUBBLES.   OIL-GLOBULES.

Once for all, it may be mentioned that no fresh animal tissue is to be examined in a dry state; it is always to be placed in water or some other harmless bland fluid.

When powders or crystalline substances are examined, they also must be diffused through water, or some liquid in which they are insoluble. Fluids, on the other hand, such as milk, blood, or urine, may be examined directly, all that is required being to place a drop on the slide, and then put on the covering glass. In order to make a good observation, only one stratum of corpuscles, or globules, or particles of any kind should be in the field of the microscope at once. If the fluid is viscid and thick, a drop of water or other liquid ought to be added, for it may be laid down as a great general principle, in all microscopic inquiries, that the more divided the objects under examination are, the clearer will be the impression they produce on the mind. If, for example, in the examination of blood, the corpuscles are heaped and crowded together, the investigator is totally unable to appreciate their true size, shape, or appearance; whereas, if there are only a dozen or so in the field, there will be no difficulty in obtaining a distinct view of each one.

So it is with all objects, the fewer in the field the better.

Sometimes it is desirable to examine a sediment at the bottom of a fluid. To do this, it is only necessary to take one of the dipping tubes, place the fore-finger over one end, and then pass the other through the fluid to the necessary depth; while the finger is kept over the end, the tube will remain empty, but as soon as it is lifted up the particles at the other end of the tube rush in with some force; when enough is obtained the finger must be again placed over the upper end, and the tube withdrawn; the substance required for examination may now be easily transferred to the centre of the slide and covered in the usual way.

In this manner minute organisms may be captured from the fluids in which they float, or even taken, when necessary, from under the microscope.

Focussing.—The object being ready for examination, the slide may be placed on the stage in such a way that the preparation may be as nearly as possible under the centre of the objective. In focussing the instrument, the great point to be attended to is to avoid a collision between the object-glass and the preparation, to the detriment of one or both. To prevent this accident, and to save time, lower the tube with the coarse adjustment, *without looking into the instrument*, until the point of the objective is near to the slide—in using the high powers it should as nearly as possible touch the thin covering glass—then, looking into the eye-piece, the tube is to be *raised* slowly until the object comes into view. This done, the focus can be readily corrected by means of the fine adjustment.

When the preparation is very minute, it will always be best to find it first with a low power, and afterwards to screw on a higher one. The lower objective should, in most instances, be used first, to obtain a general idea of the preparation before it is attempted to examine the details.

The adjustment of the focus is of the first importance in microscopic observations, and many erroneous conclusions may be traced to its incorrect performance. It is only by experience that the student can acquire the requisite knowledge and tact which will enable him to discriminate the true from false appearances.

It must be borne in mind that the higher powers are more difficult to use, and that with them the chances of error are much greater than when the lower objectives are employed; it is, consequently, best to select the lowest power that will render the details of a structure apparent, rather than to attempt the amplication of an object unnecessarily.

Preserving and Mounting objects for the Microscope.

PRESERVATIVE FLUIDS.

1. Canada Balsam . . . . 2 parts
   Benzole . . . . 1 part

2. Solution of Arsenic (1 grain to distilled water 1
   ounce) . . . . 1 part
   Strong Glycerine . . . . 1 part

3. Isinglass dissolved in hot distilled water . . 1 part
   Pure glycerine . . . . 1 part

   This forms a jelly which becomes fluid on the application of slight heat.

4. Rectified spirit . . . . 1 part
   Distilled water . . . . 2 parts

5. Creosote or carbolic acid . . . . . 1 part
   Distilled water . . . . 50 parts

6. Wood Naphtha . . . . 1 part
   Distilled water . . . . 10 parts

7. *Goadby's Fluid*—
   Bay salt . . . . 1 oz.
   Alum . . . . 2 dr.
   Corrosive sublimate . . . . . . 1 gr.
   Distilled water (boiling) . . . . . 1 pint

COLOURING FLUIDS.

In order to render certain parts distinct which were not previously apparent, tissues may be coloured by imbibition.

By this process the nuclei and granular matters are coloured more deeply than the other parts of tissues.

Carmine Fluid, which is one of the best, is made by dissolving pure carmine in ammonia and afterwards diluting with water, until a rose-colour is obtained; the mixture should be filtered. In this the tissue may be soaked for twenty-four hours, dried, and mounted in Canada balsam; or the ammoniacal solution of carmine may be added to strong glycerine until the required depth of colour is produced, and the specimen macerated in it for two or three days. It may then be mounted in glycerine, or washed in a weak solution of acetic acid, dried, and afterwards put up in Canada balsam.

A solution of Magenta may be employed for the same purpose.

Solution of Nitrate of Silver—one grain to an ounce of distilled water—is useful for colouring delicate tubes, outlines of cells, and fine connective tissue. The specimen must be soaked from a few hours to two or three days, according to the depth of colour required, after which it is to be washed, and exposed to the light until sufficiently dark in colour.

Certain tissues, such as bone, tooth, horn, and shell, require no further preservative than is afforded by the application of a covering glass, rendered permanent by the employment of some adhesive material to its edges to fasten it to the slide. Gum and paper, sealing-wax, varnish, gold size, or Brunswick black, are all available for this purpose.

There are even some soft textures which may be dried on a slide and covered in the same way, or they may be mounted in Canada balsam fluid, a drop being placed on the specimen and the covering glass applied, and subjected to slight pressure until the balsam is freed from air-bubbles and is sufficiently hardened to fix the covering glass in its position. Other tissues, which will bear being soaked in turpentine, may be mounted in the balsam fluid without any further preparation.

Most tissues, however, in the moist state, require to be kept in some fluid medium (see Nos. 2, 4, 5, 6 or 7), which will prevent their decomposition without impairing their character. For the most delicate structures, to be examined under the highest powers, strong glycerine is the best medium that can be employed, both for the preparation and also for the preservation of the object. Spirit and water (No. 4, p. 22), or a weak solution of creosote (No. 5, p. 22), or camphor water, answer very well for vegetable tissues. Goadby's fluid (No. 7, p. 22) has high preservative properties, but coagulates albuminous matters.

Glycerine may be used with almost any other medium, either in the formation of preservative fluids, or as a vehicle for all chemical tests used in microscopic examinations. In very delicate dissections glycerine will be found superior to serum, white of egg, syrup, or water.

All fluids should be filtered before being employed to put up objects, and, if possible, the specimen should be soaked for some hours in the fluid in which it is intended to be mounted.

To illustrate the steps of the process of putting up an object permanently, we will suppose the observer to have placed a drop of one of the preservative fluids on the glass slide, instead of a drop of water, the section or the fragment of tissue is to be put in this in the ordinary way, teazed out, if necessary, covered with the thin glass, just as in an ordinary examination of any portion of structure. The specimen is now to be put under the microscope, and if it prove to be a good one, it is only necessary to render it permanent by painting a little gold size round the covering glass, so as to cement it on to the slide. Several successive coats of size will be required at intervals of not less than 24 hours, that one coat may be quite dry before another is applied.

In the event of a specimen prepared in water in the usual way happening to prove so good that the observer desires to preserve it, all that he has to do is to replace the water by one of the preservative fluids, without removing the covering glass. In order to effect this, he must drop a little of the solution at one point of the edge of the cover, and apply a piece of blotting paper at the opposite point, thus drawing off the water from one side, and leaving room for the preserving medium to pass under, by capillary attraction, at the other. The blotting paper must be afterwards used to remove all moisture around the covering glass, before the gold size is painted round its edge. When a sufficient number of coats of gold size have been employed to cement the cover down firmly, the appearance may be improved by a subsequent coating of Brunswick black; this, however, is not necessary to the success of the process, which is to be conducted in the same manner, whether the object be mounted on the flat surface of a common glass slide, or placed in any of the numerous cells adapted for holding preparations which require a larger amount of fluid than could be retained upon the surface of the ordinary slide.

INJECTED PREPARATIONS.

The art of filling the vessels of a tissue with some coloured fluid to represent the blood is at once a very important and difficult one. A natural injection of the capillaries may sometimes be obtained by ligatures; the plan, however, is only applicable to membranous structures, such as intestines or stomach. Artificial injection consists in forcing a fluid into the vessels of a part or the whole of the circulatory system.

In the old method the injected medium was wax, fat, or gelatine, rendered fluid by heat and coloured with vermilion, chromate of lead, Prussian blue, or white carbonate of lead. This plan, distinguished by the term opaque injection, is now almost entirely abandoned, having been superseded by the more recent method of transparent injection, in which a transparent base is coloured by a pigment held in perfect solution, or, at least, so distributed as not to interfere with the passage of light to any considerable extent. For the purpose pure gelatine, coloured with carmine, is generally used; but for ordinary purposes the preparations recommended by Dr. Beale (one a red, and the other a blue fluid) answer very well. They have the advantage of being easily prepared, and may be always kept ready for use.

Red Fluid for Transparent Injection.

Carmine	5 gr.
Glycerine, with eight or ten drops of acetic acid	½ oz.
Glycerine	1 oz.
Alcohol	2 dr.
Water	6 dr.
Ammonia, a few drops	

Mix the carmine with a few drops of water, and add five drops of ammonia, next pour in half an ounce of glycerine and agitate. The acid glycerine is then to be gradually added, frequently shaking the bottle during the mixing. If the mixture is not decidedly acid to blue litmus paper, add a few drops of acetic acid to the remaining glycerine and mix it with the fluid. Lastly, mix the alcohol and water and pour them in gradually, agitating the bottle frequently.

Blue Fluid for Transparent Injection.

Glycerine	2 oz.
Wood naphtha or pyroacetic spirit	1½ dr.
Spirit of wine	1 oz.
Ferrocyanide of potassium	12 gr.
Tincture of sesquichloride of iron	1 dr.
Water	3 oz.

Dissolve the ferrocyanide of potassium in one ounce of glycerine, add the sesquichloride of iron to the other ounce of glycerine; pour the iron solution gradually into the solution of ferrocyanide, and agitate well. Then add the naphtha mixed with the spirit, and afterwards the water very gradually, shaking the mixture constantly.

Apparatus required for injecting are:—a proper syringe, capable of holding about two ounces of fluid, pipes with stopcocks, scalpels, scissors, forceps, bull's-nose forceps, wash-bottle, and needle and thread for passing under a vessel to tie in the nozzle of the pipe.

Structures are generally injected by the arteries, the fluid being introduced by a moderate and continued pressure upon the piston of the syringe until it flows out of the larger veins. A small animal, as a frog or mouse, may be injected by the aorta, or an eye or kidney by the artery, by the student as a commencement.

After the injection is complete the specimen may be immersed in strong glycerine or absolute alcohol before being dissected.

The best time for injection is immediately after death.*

---

* Small animals may be destroyed for dissection by putting them for a few minutes under a bell-glass with a little chloroform.

Mammalian animals and also reptiles are best injected by the aorta.

Mollusca by making an opening through the integument and putting in a pipe.

Fishes may be injected by cutting off the tail and introducing a pipe into the vessel immediately beneath the spine.

For insects a pipe may be placed in the abdominal cavity, and the vessels filled from it.

Portions of organs are to be filled from one artery, the rest being tied. Lymphatics and lacteals are most readily injected after the blood-vessels have been filled with water.

ELEMENTARY TISSUES.

CELLS—THEIR NATURE AND FUNCTION.

NOTWITHSTANDING the apparent diversity in the structure of the various tissues of which animals and vegetables are constituted, recent microscopic research has demonstrated that all textures originate from cells.

The ultimate fibres of muscles are formed of corpuscles arranged in rows; the soft tissue of the liver, and the hard texture of horn are equally constituted of cells; even the seemingly homogeneous filaments of fibrous tissue are associated with occasional nuclei which bear testimony to their cellular origin. In fine, from the first trace of embryonic life to the cessation of animal existence, all the marvellous processes of vitality—the origin, development, reproduction, and decay of the organism—are dependent upon the development and metamorphosis of cells, differing from each other in form and function, but uniformly constituted of a membrane termed the cell wall, so arranged as to form a sac capable of enclosing both fluid and solid contents.

In the interior of many cells is a small body of granular structure named the nucleus. Within the nucleus there are to be discerned occasionally one or more smaller bodies called nucleoli.

The illustration represents simple cells from the epidermis; 1 and 2 without nucleus, 3 with nucleus, 4 shows nucleus and nucleolus.

Fig. 16.

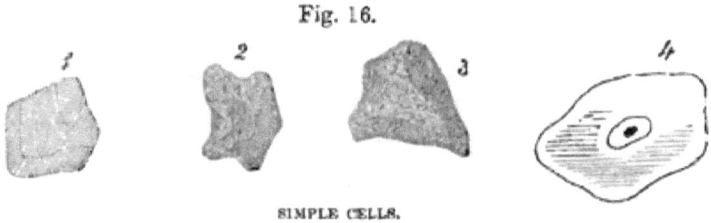

SIMPLE CELLS.

Many vegetable cells have within them a second cell called by Mohl the 'Primordial Utricle.' The wall of the outer cell is sometimes quite separate from that of the Primordial Utricle, but occasionally they are so closely connected, as to require the aid of chemical agents (alcohol or hydrochloric acid) to render them apparent.

In cartilage cells a Primordial Utricle may often be distinguished. The following woodcut represents three cells from a piece of cartilage showing the Primordial Utricle in three separate states.

Fig. 17.

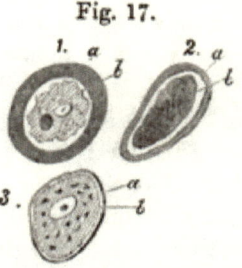

Three human cartilage-cells, magnified 350 times. 1. From the epiglottis, readily separable, with somewhat shrunk primordial utricle. 2. From an articular cartilage, with strongly contracted primordial utricle. 3. From an ossifying cartilage, with unaltered primordial utricle, the two latter cells with thin cartilage capsule. *a*. Cartilage-capsule. *b*. Primordial utricle, with the cell contents and nucleus, which in 2 is concealed.

Preparation.—The tissues of young plants, or algæ, or specimens of cartilage, may be selected for examination. Minute fragments of either object ought to be placed in a drop of water on the slide and carefully teazed out for the purpose of obtaining the cells separated from their connecting tissues in order to examine them singly; if possible, no more than two or three should be in the field of the microscope at the same time.

Cells, especially those of plants, are often connected by an intercellular substance so closely allied to the cell wall as not to be distinguishable from it even by the aid of chemical reagents.

Cells vary much in size, and, as a rule, vegetable cells are larger than those of animals.

In shape, cells may be spherical, oblong, polygonal, stellate, or fusiform.

Cells may coalesce with adjoining cells, and by communicating with them, form tubes; or they may undergo still further modifications and form fibres, bands or spiral vessels.

They may be spread out so as to form a membrane, either being immediately united by their edges or connected together by some intervening tissue.

Origin of Cells.—1. Cells may arise from a formative fluid, derived from the blood, called the 'blastema.'

In the blastema minute granules, termed cytoblasts, appear; a number of which granules becoming grouped together form the nucleus, round which the cell membrane is afterwards developed by the coalescence of a series of other granules. The nucleus may either remain in the centre, or become attached to some part of the cell wall.

2. Cells may arise within parent cells by an endogenous process of development.

The first step is the formation of several nuclei by the breaking up of the original nucleus; the second step is the subdivision of the cell's contents into as many portions as there are nuclei, each portion enclosing a nucleus; lastly a new cell wall forms round each part and completes the process. In this way a number of cells may be enclosed within one common investing membrane, as seen in the cleavage of the yolk of the ova of some parasites.

Fig. 18.

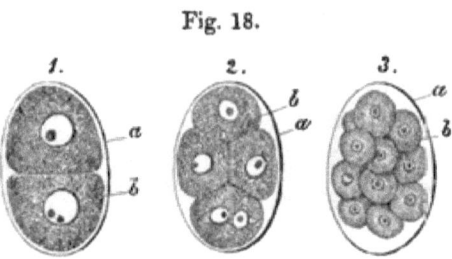

Three ova of Ascaris Nigrovenosa. 1. From the second, 2. from the third, and 3. from the fifth stage of cleavage, with two, four, and sixteen globules. *a.* External ovula or envelope. *b.* Cleavage globules. In 1, the nucleus of the lower globule, containing two nucleoli. In 2, the lowermost globule, two nuclei.

Another form of endogenous development occurs in cartilage, commencing by the division of the nucleus into two parts, followed by the passage of a partition from the cell wall of the parent separating the cell into two parts, each part enclosing a nucleus. It often happens that the parent cell wall remains entire for some time, and the same process is repeated until one original cell may contain two or three generations, as the next illustration shows.

Fig. 19.

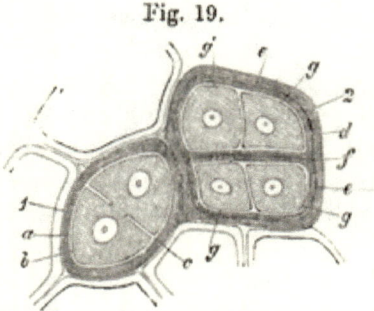

Cartilage-cells of a full-grown Tadpole, semidiagrammatical figure. 1. A mother cell, whose primordial utricle is in the act of dividing. *a.* Its thick secondary membrane or cartilage capsule. *b.* Primordial utricle, enclosing the cell contents with the nucleus. *c.* Place where it is constricted (not observed). 2. A mother cell with two generations. *d.* Outer cell membrane of the mother cell. *e.* Outer cell membranes of the cartilage capsules of the secondary mother cells enclosed by them, which, with *f*, form a double partition through the chief mother cell. *g g.* Secondary cells.

3. New cells may be produced by the process of division or budding. In this, as in the other methods of multiplication, the nucleus is seen to divide first, then the cell becomes constricted in the middle, and lastly splits up into two; this process of development is observed in free cells floating in fluids—the blood of the embryo chick for example.

The following illustration represents the multiplication of the blood globule of the embryo chick by division.

Fig. 20.

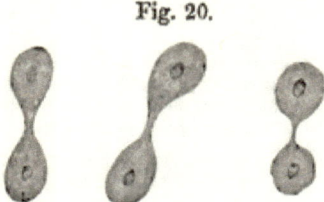

Blood globules of chick, in the act of division.—Magnified 350 diameters.

In the juices of glands the new cells appear to be formed as buds upon the original cells, and ultimately to separate from them, and then give origin to others in the same manner.

FUNCTIONS OF CELLS.

On the Function of Cells.—Simple Cells forming a Covering.—
A good example is found in the delicate tissue covering the inner layers of the onion; in order to examine it, remove some of the outer portion of an onion, and then with the point of a knife peel off a piece of the fine membrane from one of the inner layers, place it in a drop of water on the slide, apply the covering glass as directed, and view the object under a low power.

If the preparation has been properly made, the tissue will appear as a single layer of fine oblong transparent cells, united by their sides and edges, with here and there a nucleus in their walls.

Fig. 21.

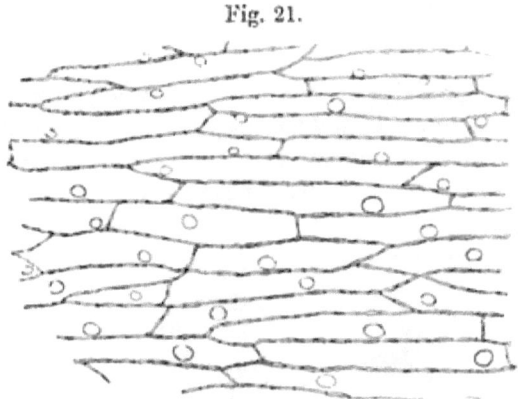

Membrane from Inner Layers of an Onion.—Low power.

Cells performing the Office of a Store-house.—In the potato and other bulbous tubers, there are cells in which are stored up numerous granules of starch.

Preparation.—Obtain, if possible, the fresh root of the *Iris Germanica* for examination.

Make a very thin transverse section with the razor, place it on the slide in a drop of water, put on the covering glass, and examine with the high power.

The large polygonal cells appear filled with starch granules, collected in different parts of their interior, and a drop of a dilute solution of the tincture of iodine, applied to the edge of the covering glass, will soon render the starch granules more distinct, by forming with them the blue iodide of amylum.

Fig. 22.

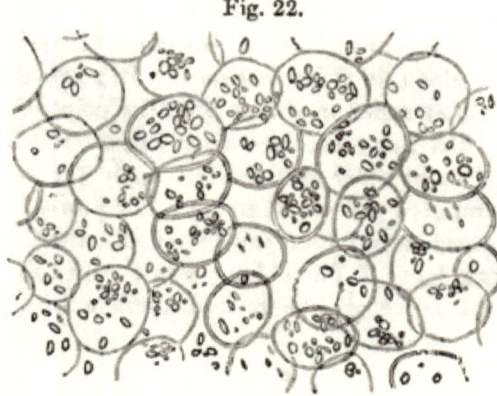

Section of Root of Iris Germanica.—Low power.

The starch granules of the potato vary much in size; many of them are very large and furnish good examples for examination. Either a fine section of a potato, or a little of the pulp scraped off, is to be placed in a drop of water on the slide, covered and examined under a high power. The granules will appear as shown in the drawing when highly magnified.

Fig. 23.

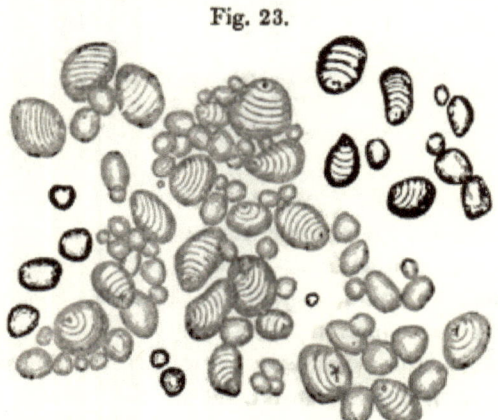

Starch Granules from a Potato.—High power.

A minute spot, called the hilum, will be noticed in each granule, marking the point of its attachment, in its newly developed state, to the wall of the cavity in which it occurs. Round the hilum there are well-marked concentric lines.

Instead of nutritive material, like starch, cells may contain colouring matter.

Such cells are found in the integument, in the hairs, in the choroid coat of the eye, and in many other textures.

In order to obtain specimens of pigment cells take an eye of a bullock or other animal, cut it open and allow the humours to escape, peel off the retina, and with the point of the scalpel slightly scrape the black part of the choroid coat, avoiding the metallic coloured part; place the black matter so collected in a little water on the slide, taking care that a very small quantity only shall be distributed through the fluid, apply the covering glass in the usual way, and examine the object with a high power.

A number of hexagonal cells will be seen, filled with pigment granules, and having often a light-coloured nucleus.

The drawing represents the cells distinctly, but their outlines are seldom so regular as depicted.

Fig. 24.

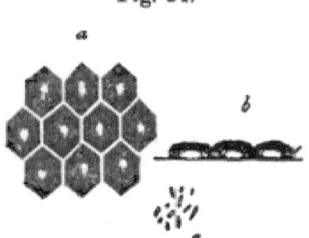

Cells of the black pigment of man. *a.* Seen from the surface. *b.* From the side. *c.* Pigment granules.—High power.

The depth of the colour of the pigment cell depends upon the number of granules which it contains.

Complex Cells.—Cells may have more than one office to perform. For example, the cuticular coverings of many plants are furnished with stomata, or breathing pores. Stomata do not exist in plants which grow under water; but in those whose leaves float on the surface, there are stomata on the side exposed to the air, and none on that in contact with the water. In some instances, as many as 160,000 pores have been found in a square inch of surface.

A leaf of the *Iris Germanica* should be selected and the thin transparent cuticular layer, peeled from the surface, placed in a drop of water on the slide, covered and examined with both low and high powers.

Fig. 25.

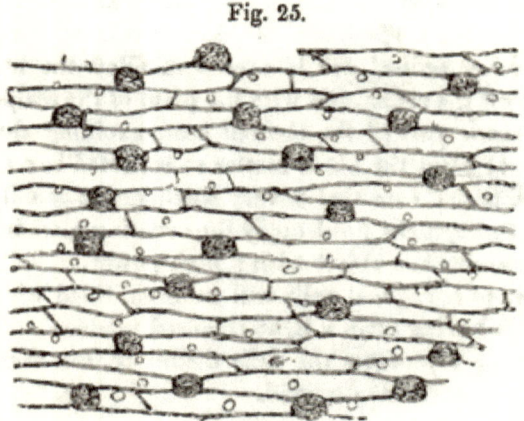

Cuticle of Iris, showing Stomata.—Low power.

The stomata consist of two kidney-shaped cells (*a*) having granular contents. Each pore opens into an air cavity.

Cells with Moving Contents.—In many water plants, the cell contents are in a state of constant activity, giving rise to the appearance of a circulation, from which, however, the action is entirely distinct. In the *Vallisneria spiralis* (a native of Southern Europe) or the *Nitella flexilis*, the process is most easily examined.

Preparation of Vallisneria.—A portion of the moist fresh plant being laid on the thumb-nail of the left hand, thin longitudinal sections should be made with the razor, and immediately placed in warm water (80° to 100° Fahr.).

After a short time a specimen may be transferred with a drop of the water to the slide, covered quickly and examined by the high power. Should the movements not be apparent, the slide should be slightly warmed over a spirit lamp. When the preparation is successful, the chlorophyll granules will be seen moving down one side and up the other of the cell wall, never passing in opposite directions on the same side.

In some places, the granules will be observed moving round a centre, or passing in a continuous stream through long tubes.

The movements will occasionally continue when the plant is putrid.

Nitella requires no preparation beyond placing a small piece of the plant on the slide in a drop of water, covering it, and examining the object under a high power.

In a small zoophyte trough, *Nitella* may be kept growing for some time, always ready for observation.

Epithelial Cells.

All free surfaces of the body, both internal and external, are covered by a layer of cells termed epithelium. Their structure is everywhere the same, although in different parts of the body they present certain modifications in their arrangement.

Epithelial cells are nucleated, and always joined by their surfaces or edges, without the intervention of any connective tissue.

There are four essential varieties:—1. Tesselated; 2. Columnar; 3. Spheroidal; 4. Ciliated. In all these forms the nucleus remains remarkably uniform in its characters. It is round or oval, and flattened $\frac{1}{6000}$ to $\frac{1}{4000}$ in diameter, insoluble in acetic acid, colourless or slightly reddish in tint, and usually contains one or more nucleoli, with a few irregular scattered granules.

TESSELATED EPITHELIUM.—Scaly, lamellar, squamous, tabular, pavement, or flattened epithelium. This form of epithelium is the most common. It is found in single layers lining serous cavities, on many parts of the mucous membrane, and in the interior of blood-vessels and ducts.

Upon the surface of the skin it occurs in superimposed layers, forming the 'stratified' epidermis.

To obtain specimens of scaly epithelium, it is only necessary to scrape the lining membrane of the cheek with a scalpel, and place the fluid collected in a drop of water on the slide, cover with the thin glass in the usual way, and examine under the high power.

The drawing represents what the observer should see—a number of delicate flattened cells of polygonal form, containing a few granules, and generally a nucleus and nucleoli.

HISTOLOGY.

Fig. 26.

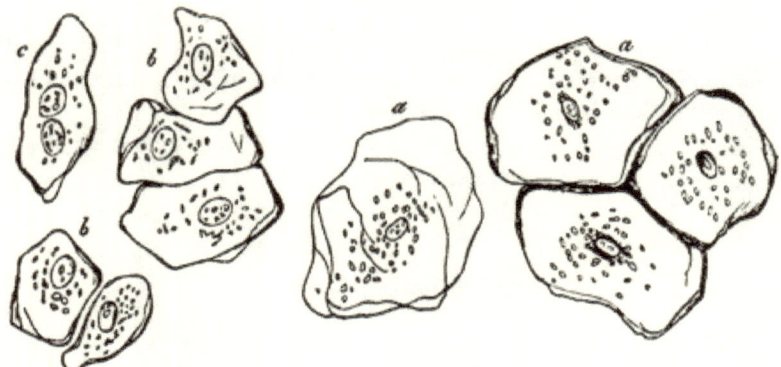

Epithelial cells in the oral cavity of man. *a.* Large. *b.* Middle sized. *c.* The same with two nuclei.—Magnified 350 times.

Scaly epithelium, as will afterwards be pointed out, is the elementary constituent of horn, nail, and hair.

COLUMNAR EPITHELIUM.—This variety exists upon the mucous membrane of the stomach, on the villi of the intestines, on the membrane lining the urethra, and most other canals.

It is set on to the surface perpendicularly, and may be detached in rows.

Preparation.—A portion of the stomach or intestines of any animal—sheep or pig, for instance—during the digestive process will furnish examples of columnar epithelium.

Scrape the surface of a portion of the mucous membrane of the stomach or intestine lightly with a scalpel, and place the matter thus obtained in a drop of water on the slide, apply the cover, and examine with the high power. The cells will appear as in the illustration if the preparation is successful.

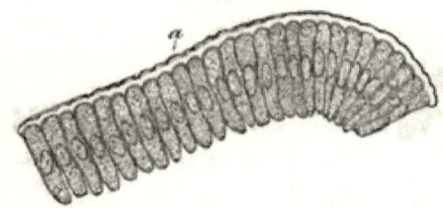

Epithelium of the intestinal villi of the rabbit.— Magnified 300 times.

SPHEROIDAL EPITHELIUM is found in the bladder, ureters, pelvis of the kidney, and in the ducts of the secreting glands.

Preparation.—Specimens may be obtained by taking a piece of the mucous membrane of the bladder, and, scraping the surface lightly with a scalpel, the collected matter must be placed in a drop of water on the slide and examined as a covered object with the high power.

The cells are, for the most part, circular, although some are a little flattened at the sides where they are in contact with each other.

Fig. 28.

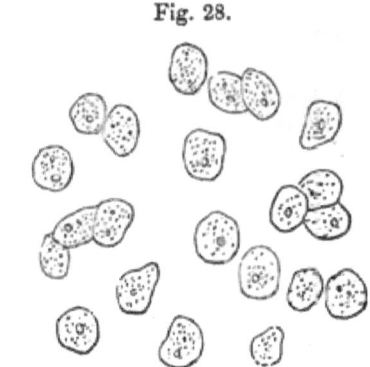

Spheroidal Epithelium from Human Bladder.—Magnified 250 diameters.

CILIATED EPITHELIUM is characterised by the presence of fine hair-like filaments (cilia) attached to the free surfaces of the cells.

During life, and for some time after death, these hair-like bodies are in constant waving motion. They all move in one direction, and rhythmically, which gives rise to the appearance of a succession of undulations. Ciliated epithelium is found in the air-passages extending to the bronchial tubes, in the uterus and fallopian tubes, in the tympanum of the ear, and Eustachian tubes, and in the ventricles of the brain; in fact, wherever it is necessary to urge on a secretion by mechanical means, ciliated epithelium exists and performs the office of the scavenger's broom.

Specimens for examination are easily obtained, and with a little care will show the characteristic motion for some considerable time.

Preparation.—Select a perfectly fresh mussel, open its shell, and remove, with the point of a pair of small scissors, a small portion of the yellow fringe or mantle, which will be easily seen lining each half of the shell; place it, with a drop of *salt* water, on the slide, cover with the thin glass without pressure, and examine with the high power.

The cilia will be seen in a state of active lashing movement, and if the specimen has been carefully prepared this movement

Fig. 29.

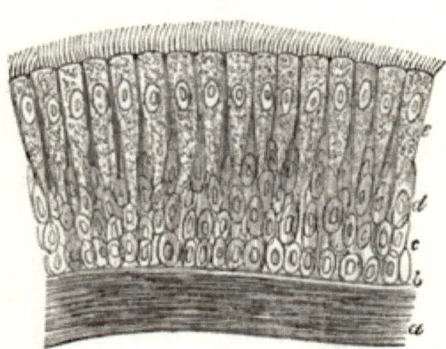

Ciliated epithelium, from the human trachea; magnified 350 times. *a.* Innermost part of the elastic longitudinal fibres. *b.* Homogeneous innermost layer of the mucous membrane. *c.* Deepest round cells. *d.* Middle elongated. *e.* Superficial, bearing cilia.

will continue for a long time, during which small particles of dust or other foreign matter that may have accidentally got between the slide and the covering glass will be seen driven along by the ciliary movements.

Lymph, Chyle, and Blood.

These fluids present, in one respect, a physical uniformity of composition, consisting of certain characteristic corpuscles, distributed through a fluid medium.

LYMPH may be obtained from a lymphatic vessel of any large animal, or, better still, from the thoracic duct of an animal that has fasted for some hours before death, which is ordinarily the case with animals slaughtered by the butcher.

By simply puncturing the vessel with a needle, a drop of lymph may be caused to exude; the slide should be immediately placed on the drop, which will thus be transferred to it, the covering-glass, previously breathed upon in the usual way, must then be applied, and the object examined under a high power.

The lymph corpuscles will be seen as depicted in the drawing, agreeing in their general characters with the white corpuscles of the blood, except that they vary much in size, and sometimes contain smaller cells or nuclei embedded among the granules.

CHYLE contains, beside the corpuscles of lymph, a quantity of minute granules, forming the molecular base a (fig. 30), which gives to the fluid its white colour. There are, besides, oil globules, free nuclei, and sometimes a few red blood discs.

Fig. 30.

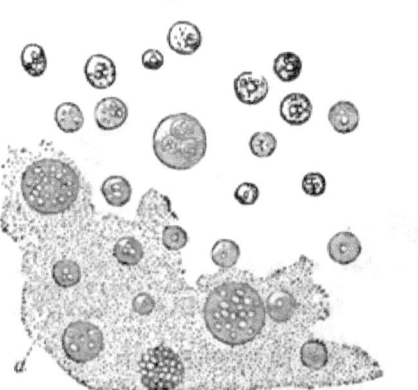

CHYLE FROM LACTEALS.

Chyle may be obtained for examination from the lacteal vessels of an animal in full digestion; a drop may be taken in the same manner as lymph, and examined, without the addition of water, under a high power.

Corpuscles identical with those of lymph and chyle are easily obtained for examination by squeezing a little of the juice from a lymphatic gland.

BLOOD CELLS vary considerably in mammals, birds, reptiles, and fishes.

In order to obtain a specimen of human blood cells, it is only necessary to bind the finger a short distance from the tip, with the corner of a handkerchief, in order that it may become congested, and a slight puncture with a fine needle will cause a drop of blood to exude. As soon as it appears touch it lightly with the glass slide, breathe on the thin covering glass, and directly apply it with sufficient pressure to squeeze out the greater part of the drop of blood, so that only a thin nearly transparent film remains.

Examine with the high power, and if too many corpuscles are present, the excess must be got rid of by a repetition of the pressure. To see the blood corpuscles well, not more than twenty or thirty should be in the field at one time.

In the illustration the blood corpuscles are seen in the various positions, in which they are frequently to be observed.

Fig. 31.

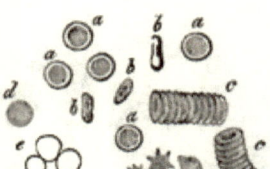

Human blood-globules. *a.* Seen from the surface. *b.* From the side. *c.* United in rouleaux. *d.* Rendered spherical by water. *e.* Decolorised by the same. *f.* Blood-globules shrunk by evaporation.

The red discs of human blood are distinguished by their clearly defined outlines and dark centres. Each disc is biconcave, and hence the whole surface cannot be focussed at one time; thus, when the circumference is in light the centre, being a concavity, is dark, but by bringing the object-glass nearer to the object, the concavity is brought into focus and becomes light, leaving now the margin dark.

By moving the fine adjustment in both directions, it will be seen that the dark parts become light, and the light portions dark, as each is alternately in and out of focus.

This experiment should be repeated until the changes are perfectly understood.

It occasionally happens that the corpuscles adhere together by their flat surfaces, and look like rolls of coins, as in the illustration; at other times, even in the same person, they manifest no such tendency, and are irregularly scattered throughout the field in the liquor sanguinis.

A single disc presents the following characters: its form is circular in all mammalia except the camel, dromedary, and llama, which have oval blood discs; in profile it is biconcave; the investing membrane is smooth, slippery, and elastic, capable of gliding readily along the vessels and altering its form to suit the calibre of the canals through which it is compelled to pass in the course of the circulation.*

There is no trace of a nucleus in the blood discs of any of the adult mammalia.

In size the blood discs bear no proportion to the bulk of the animal in whose vessels they circulate, as the following table will show.

	Inch.		Inch.
Musk Deer	$\frac{1}{12000}$	Ape	$\frac{1}{3400}$
Goat	$\frac{1}{8400}$	Man	$\frac{1}{3700}$
Red Deer	$\frac{1}{5000}$	Elephant	$\frac{1}{2700}$
Horse	$\frac{1}{4000}$	Fish (eel)	$\frac{1}{2000}$
Sheep	$\frac{1}{4500}$	Frog	$\frac{1}{1600}$
Cat	$\frac{1}{4400}$	Salamander	$\frac{1}{700}$
Ox	$\frac{1}{4200}$	Proteus	$\frac{1}{400}$
Fox	$\frac{1}{4100}$	Cryptobranchus Japonicus	$\frac{1}{800}$ by $\frac{1}{350}$
Wolf	$\frac{1}{3000}$		
Dog	$\frac{1}{3500}$		

The amount of the red corpuscles can only be determined approximatively.

Welcker found, by a method of counting, in one cubic millimetre:

5,000,000 in men.
4,500,000 in women.

---

* Dr. Roberts concludes that the blood disc like the vegetable cell has an interior vesicle containing the colouring matter; analogous to the primordial utricle, in consequence of water occasionally causing the outer cell-wall to give way at one spot and allow a small vesicular-looking body to project, which, he thinks, is a portion of the inner sac. This appearance is very common in blood passed from the kidney in cases of albuminuria.

During pregnancy, amenorrhea, chlorosis, and anæmia, and also after venesection or abstinence, the number of red corpuscles is considerably diminished.

In venous blood they are more numerous than in arterial, and the blood of the hepatic veins is said to contain the greatest quantity.

White corpuscles differ materially from the red discs; in man they are larger than the red, and have an irregular or finely granular surface. By the addition of acetic acid a nucleus is rendered visible, and, according to Mr. Wharton Jones, if strong acid be employed the nucleus is split up into three or four parts, a delicate envelope at the same time coming into view, ultimately dissolving, and setting the nuclei free.

The proportion, in health, of white corpuscles to the red is said to be in man as 1 in 300 or 400.

Under various conditions, the relation may vary even to the extent of 1 in 1,000. The number is increased by fatty food, during pregnancy, and after venesection.

Kölliker remarks, that in the horse, 'after enormous evacuation of blood up to fifty pounds,' the white corpuscles become almost as abundant as the red.

In the splenic vein they are very numerous (1 in 60), while the splenic artery, according to Hirst, contains only 1 in 2,200.

In order to examine the white corpuscles, the easiest method, when they are not sufficiently numerous to be readily distinguished among the red discs, is to add a drop of water to the specimen of blood, while under the microscope; in a short time, the red discs will become indistinct, and leave the white corpuscles in possession of the field, as seen in the accompanying drawing.

Fig. 32.

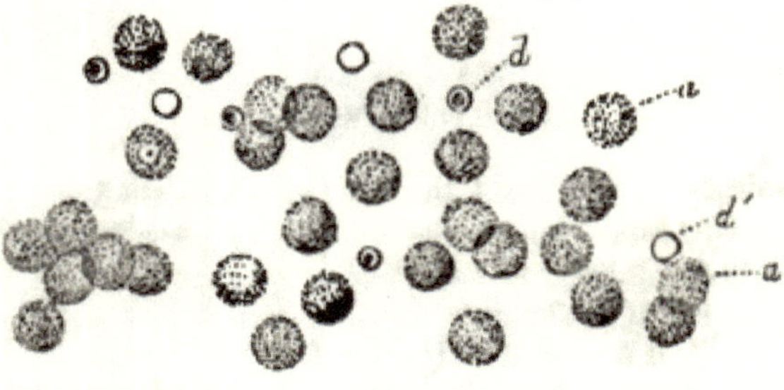

a. White corpuscles of human blood.   *d.* Red corpuscles.—High power

Besides the red and white blood corpuscles, there are certain occasional elements observed; for example, cells enclosing blood corpuscles in the blood from the spleen; pigment cells, and colourless granular cells in diseased splenic blood; concentric bodies, three or four times larger than the blood discs; fibrinous coagula; caudate cells; and free granules.

The plasma of the blood exuded from inflamed surfaces, or obtained from the 'buffy coat' of blood, and allowed to coagulate under the microscope, is seen to consist of a multitude of filaments interwoven, as in a piece of felt, more or less intermixed with corpuscules and fine granules.

To preserve specimens of blood corpuscles, dip a fine needle in the drop of blood as it exudes, and draw thin lines across the slide, allow the preparation to dry, and it will remain perfect without any covering glass for a long time.

Sections of blood discs are made by drawing lines of blood with a needle over the slide, allowing the specimen to dry, and then cutting the lines across in all directions with the razor. The loosened portions can then be brushed off with a small brush.

In birds the blood discs are oval in shape and possess a nucleus which is rendered apparent by acetic acid. For examination, a drop of the fluid may be taken from a fowl or any small bird by means of a slight puncture with a needle; the drop should be transferred to the slide, and prepared exactly as directed for human blood.

Under a high power the globules will appear as in the illus-

Fig. 33.

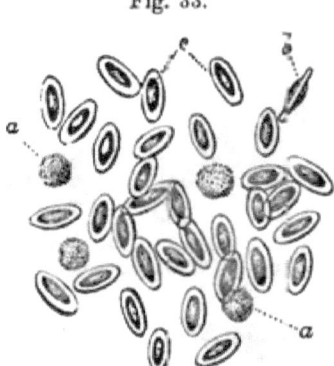

Blood discs of the fowl. *a*. White corpuscles.—High power.

tration, the white corpuscles being easily distinguished among the oval red discs by their smaller size, circular form, and granular aspect.

The blood discs of the pigeon, on drying, diminish in breadth more than in length, and thus appear to be elongated.

The blood of fishes has oval and nucleated discs, rather more pointed than those of birds.

Blood cells from the codfish or eel often become triangular after standing for some time.

The white corpuscles have the circular form and granular characters common to them in all animals, and vary but little in size; hence in animals with small red corpuscles they are larger, and in animals with large red corpuscles they are smaller than the red.

In reptiles generally the red blood discs are large oval nucleated bodies; the white corpuscles still preserving their invariable circular form and granular appearance.

For demonstration, the blood of the frog may be taken and prepared according to the directions previously given.

The corpuscles and discs will be readily seen as depicted in the drawing, if the specimen has been carefully prepared.

Fig. 34.

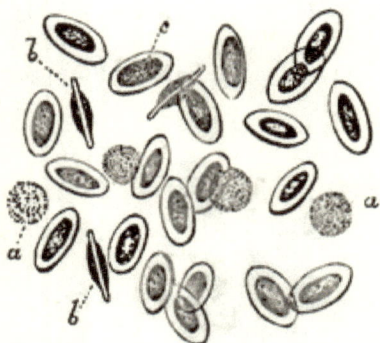

Blood discs of the frog. *c.* Nucleus. *b.* Disc seen in profile, showing a double convexity. *a.* White corpuscles.—High power.

Changes induced in the Blood Cells by Re-agents.—Water, added to a specimen of blood under the microscope, causes the red discs to become spherical and indistinct; they are, however, again rendered apparent by a drop of tincture of iodine.

Potash darkens the colour of the blood and rapidly dissolves the cell wall.

Solution of tannin (three grains to one ounce of water) added to blood causes a most peculiar change, as observed by Dr. Roberts. The cells first become spherical, and shortly a small bud projects from the outside of the cell wall; sometimes the protruded part is covered by a hood.

Fig. 35.

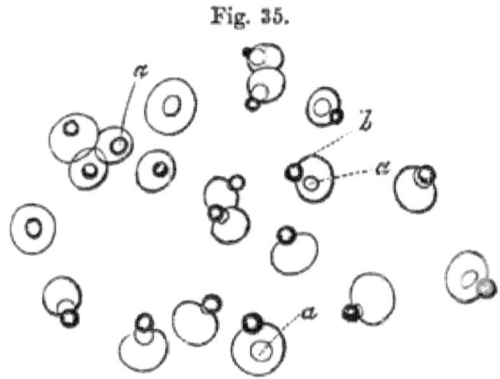

Specimen of fowl's blood after the addition of solution of tannin.
a, Nucleus. *b*, Hood.—High power.

Similar effects may be observed in the blood of birds, fishes, and reptiles, while in addition to the projection of the 'bud' from the surface, the nucleus in the interior of the cell becomes more distinct.*

Solution of salt causes the red discs to shrivel and ultimately to assume the stellate appearance seen in fig. 31.

When it becomes absolutely necessary to mix a fluid with the blood, as in the examination of dried blood, either serum, glycerine, or white of egg, must be employed instead of water for the purpose; otherwise, the change in the form of the discs, and their rapid disappearance, will render it impossible to make a proper observation.

* Identical changes occur on the addition of a drop of water to fowl's blood, followed immediately by a drop of solution of magenta.

Observation of the Blood in Circulation.—By the aid of very simple mechanical arrangements, the circulation of the blood in the smaller vessels may be readily demonstrated.

The web of the frog's foot, the wing of a bat, the tail of a small fish, or the umbilical vesicle of a newly-hatched salmon or other fish, easily obtained now artificial hatching is so common, will furnish excellent objects for the observation.

A small fish is most easily arranged, as it is only necessary to place a small pillow of cotton wool, well soaked with water, at one end of the glass slide, and then adjust a minnow or stickleback so that the head shall rest on the wool; next drop a piece of thin covering glass on the thin transparent part of the tail, and place it under a low power first, using a higher one afterwards if desirable.

The fish will generally lie quiet sufficiently long to permit a good observation to be made; but should it move, it is only the work of a moment to replace it, as no fastenings are employed.

Arteries will be at once distinguished from veins, by the direction of the current of blood, which, in arteries, runs *from* the larger *to* the smaller branches, and in the veins directly the opposite. Under a high power the red discs will be seen moving with great velocity, while the white corpuscles are rolling slowly along the sides of the vessel.

When it is necessary to make a more extended observation, the web of the frog's foot will answer best, because the part may be fixed, and the animal so placed as to suffer no injury for some considerable time.

Method of preparation.

1. Roll the frog in some wet rag and fasten it to the frog-plate or a wooden substitute, which can be easily made and answers just as well. The fastening is accomplished by twisting some tape round the body of the animal and the plate, leaving one hind-leg out of the roll of wet rag.

2. Tie pieces of thread to two claws, and stretch, but not tightly, the intervening web over the hole in the plate, or piece of wood used instead, and fasten the threads by pulling them into little niches cut in the edges of the wooden plate, or by twisting them round the pins in the brass plate.

The whole process need not occupy five minutes.

In order to fix the part under the object glass, a temporary stage of books may be built up or the ring of a retort stand may be fixed to the level of the microscope stage, to support the heaviest part of the animal, unless the stage has a clip sufficiently strong to bear the weight of the preparation.

Blood Crystals.

In addition to the elements already described crystalline forms are also to be found in the blood. These, though varying in the blood of different animals, yet preserve, in reference to each species, sufficient uniformity to render them characteristic, as shown at fig. 36, A, B, C, D.

The crystals are formed when a little blood is mixed with water on the slide, covered, and allowed to rest for a short time. Near the edge of the covering glass they are usually most distinct. The specimens should be examined under a high power.

In the next illustration (fig. 36) the characteristic forms found in the blood of several animals are depicted. In human blood the crystals are prismatic; in the blood of the squirrel, hexagonal; in the guinea-pig, tetrahedral; and in the rat and mouse octahedral. Other forms may be obtained by the aid of chemical agents, or may occur spontaneously as the result of disease.

In human blood there are three distinct crystalline forms. 1. *Hæmatin crystals, found in normal blood*, particularly from the spleen and represented at fig. 36, D. They may be always rendered apparent by the addition of a little water to blood or by agitation with æther, so as to dissolve the cell-walls of the blood corpuscles and allow the contents to escape. A drop of this mixture, placed on a slide and allowed to evaporate slowly, will furnish a number of crystals, large enough to be examined under a low power.

2. *Hæmatoidin crystals* are pathological products met with when blood is extravasated and coagulates within the body, and the liquid parts are gradually drained away, as happens in the case of old apoplectic clots. A small portion of the effused blood should be placed on a slide, covered, and examined with a high power. The crystals will be seen in the form of large plates, as represented at E.

48   HISTOLOGY.

3. *Hæmin crystals* are artificial chemical products, the result of the action of glacial acetic acid upon blood corpuscles. They are best prepared as follows: Take a little dried blood, mix it

Fig. 36.

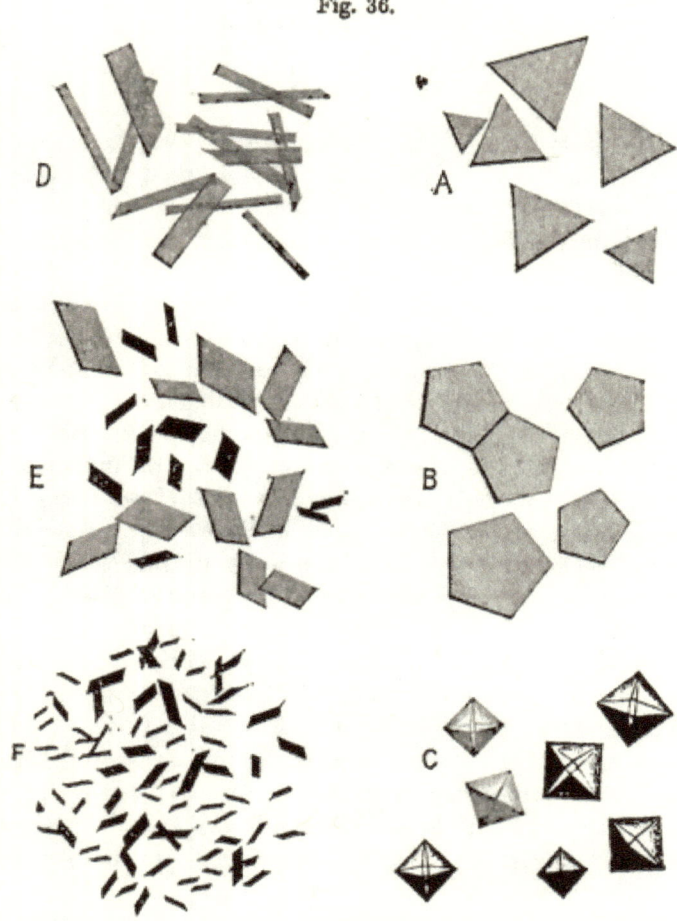

Blood Crystals. A. Trihedral crystals from blood of guinea pig. B. Pentagonal crystals from blood of squirrel. C. Octahedral crystal from blood of rat and mouse. D. Hæmatin crystals from normal human blood. E. Hæmatoidin crystals from an old apoplectic clot. F. Hæmin crystals from blood treated with acetic acid.

with an equal quantity of common salt, and boil it in a few drops of glacial acetic acid until the whole is dissolved. After the mixture has become cool, place a drop of it on a slide and examine it with a high power. A number of exceedingly minute reddish-brown crystals will be seen, as represented in the illustration (fig. 36) at F.

BASEMENT MEMBRANE.

By this term is indicated a fine homogeneous expansion, which, under the highest powers of the microscope, presents no trace of structure. It exists in certain parts of the body where epithelial cells occur, but its existence is not so general as was at one time supposed. In transverse sections of compound membranes the basement membrane is distinguishable as a limiting line only, immediately beneath the epithelial covering.

Fibrous Tissue.

There are two varieties of fibrous tissue found in the animal body. White fibrous tissue, which is flexible but not extensile; and yellow fibrous tissue, which possesses the property of elasticity as well as that of flexibility in a remarkable degree.

White fibrous tissue exists under a variety of forms in different parts of the body: as cords or tendons attached to the ends of muscles; as bands or ligaments uniting the ends of bones; as a web in fascia, spread over the surface of muscles; as a membraneous covering in the periosteum surrounding bones, and the dura mater enclosing the brain; and as capsules enveloping the more delicate parts of organs, as in the eye-ball and testicle.

White fibrous tissue is also found in the skin and other membranes, in the coats of ducts and tubes, and in all parts where strength and flexibility are necessary.

The structure of white fibrous tissue is in all situations the same, but a portion of fresh tendon will furnish the best example, and it is very easily prepared.

From a piece of fresh tendon obtained from the butcher, cut, by means of scissors, a small fragment in a longitudinal direction; place it in a drop of water on the slide, and teaze it out with the needles until it is separated into the smallest shreds; apply the covering glass in the usual way, and examine the specimen with the high power.

If the preparation is successful, the peculiar long wavy fibres will be seen as depicted below.

Fig. 37.

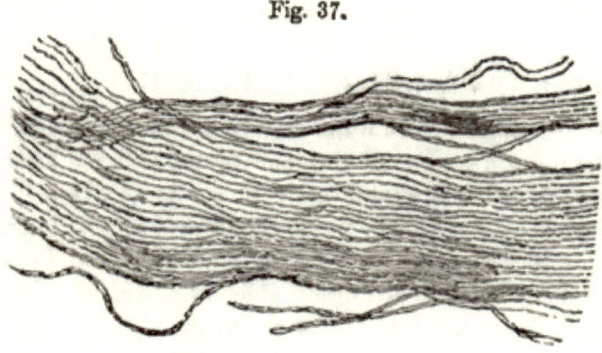

White fibrous tissue.—High power.

The filaments which are apparently homogeneous are exceedingly minute, measuring $\frac{1}{50000}$ to $\frac{1}{25000}$ of an inch in diameter; they run parallel to each other in wavy lines, but the wavy appearance is lost when the tissue is stretched.

The filaments neither anastomose, subdivide, nor interlace, although the bundles, which they unite to form, intersect each other occasionally in all directions.

White fibrous tissue is sometimes described as a solid structure, having longitudinal parallel markings or creasing, and showing a tendency to split up into fibres in the longitudinal direction to almost any extent. This view is considered to be strengthened by the fact that a drop of acetic acid, allowed to flow under the edge of the covering glass, will cause the specimen to swell up, lose its fibrous character, and ultimately become indistinct.

Transverse sections of white fibrous tissue may be made by drying a piece of tendon until it becomes sufficiently firm to be cut with a razor; very thin slices may then be shaved off, of which the best one should be selected, and placed in a drop of water on the glass slide; in a few seconds it will have resumed its natural character, and only requires to be covered with the piece of thin glass in the usual way.

Under a low power the cut ends of the fibres, and the divisions marking the bundles, with the intervening connective tissue, will be beautifully seen, but under a high power other structures will be observed, as shown in the illustration.

Fig. 38.

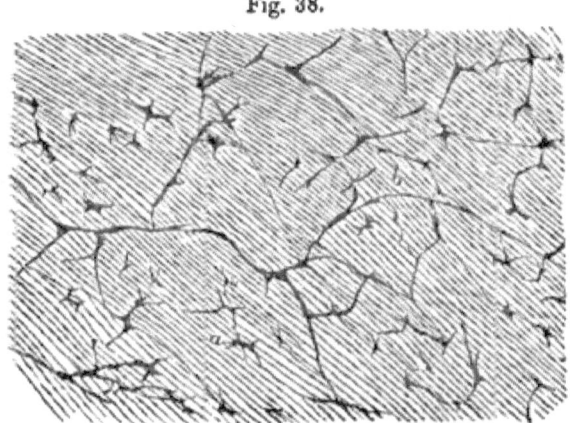

Transverse section of tendon. *a.* Corpuscles. *b.* Space occupied by areolar tissue.—High power.

The small dark points with lines radiating from them, forming a delicate network, are probably connected with the function of nutrition in the dense structure of tendon, much in the same way as the lacunæ and canaliculi of bone are concerned in the nourishment of that tissue. Virchow calls these bodies 'connective tissue corpuscles.'

Besides these corpuscles, or, more correctly, vacuities, there may sometimes be observed round or oval bodies, probably cell nuclei, apparently attached to the sides of the bundles of filaments, or lying in the interstices: they are best seen after the addition of acetic acid.

Yellow or Elastic Fibrous Tissue.—Yellow fibre is remarkably elastic and capable of considerable extension. It is found in the coats of blood-vessels; between the vertebral arches; and in quadrupeds it forms a strong elastic band, extending from the occiput to the spines of the vertebræ (*ligamentum nuchæ*), enabling the animal to support the head in a pendant position without muscular exertion.

In the ox, horse, elephant, and some others, it also exists as a firm aponeurotic expansion over the abdominal muscles.

Elastic tissue is a constituent of the skin, mucous, and serous membranes, and also of the areolar or cellular tissue.

In order to examine its structure, take a portion of the *ligamentum nuchæ* of the ox, easily obtained from the butcher, cut off a small fragment with scissors, place it in a drop of water on the slide and teaze it out well with the needles, as directed for the preparation of white fibre, cover the specimen with a piece of thin glass in the usual way, and examine it with the high power.

The appearance of the specimen should be exactly that of the illustration.

Fig. 39.

Yellow elastic tissue from *ligamentum nuchæ*.—High power.

The fibres are usually much larger than those of white fibrous tissue, although they vary considerably in size in different parts of the body from $\frac{1}{24000}$ to $\frac{1}{4000}$ of an inch in diameter; they are transparent and possess a clearly defined outline, which in some cases appears to be double. The fibres divide into two, or even three, branches, anastomose freely, and have a disposition to curl round at the broken ends.

Acetic acid produces no alteration in the structure or disposition of the fibres.

Transverse sections of yellow fibrous tissue may be made by drying a portion of the *ligamentum nuchæ*, and then cutting fine slices with a razor. The section must be placed in a little water on the slide, covered in the usual way and examined with a high power. The ends of the fibres will be seen to be hexagonal in form, giving to the specimen somewhat the appearance of a honeycomb.

Areolar Tissue, Cellular or Connective Tissue.

Areolar tissue plays a very important part in binding together the elementary constituents of the frame. It is found abundantly under the skin, and also beneath mucous and serous membranes, as well as between muscles and bloodvessels. It occupies the spaces between the various organs and parts of the body, investing them sometimes with a special sheath, or acting as a medium through which they may receive their nerves and vessels.

The minute structure of areolar tissue includes both the white and yellow elements in different proportions, according to the part of the body from whence the specimen is taken.

For the purpose of examination a fragment may be dissected from any part of an animal where the skin is loose, under the shoulder for example. The portion should be placed in a drop of water on the slide and carefully teazed out until its apparent threads are completely separated, the covering glass must next be applied and the specimen examined under a high power.

The wavy filaments of white fibrous tissue will be immediately recognised, but the yellow fibres may not be discovered without some difficulty, as they are usually very small and few in number; a drop of acetic acid allowed to flow under the edge of the covering glass will often render them apparent, by causing the white fibres to become indistinct; the yellow fibres will then be recognised by their defined outlines, by their branching and anastomosing, and by their curled ends. In some cases, a long elastic fibre is found winding round a bundle of white fibres. This peculiar appearance is most common in the areolar tissue accompanying the arteries at the base of the brain.

Adipose Tissue.

Fat is found in many situations in the animal body: in the subcutaneous areolar tissue, in the bottoms of the orbits, at the flexures of joints, surrounding the kidneys, in the mesentery, on the heart, and in the medullary cavities of bones.

True adipose tissue consists entirely of vesicles formed by a delicate homogeneous membrane enclosing the fat.

In all cases, the vesicles of adipose tissue are held together by an interlacement of fine fibres of connective tissue; although, in many instances, the vesicles are so closely arranged as to lose their circular or oval form and become polyhedral from pressure.

The fat vesicle appears to be originally furnished with a nucleus which is not often observed, as it disappears very rapidly after the fat cell is fully developed. Sometimes the fatty acids in the interior of the vesicle crystallise, giving rise to a very beautiful starlike appearance. This may often be seen in the fat of cold roast beef, but it is also to be observed in uncooked tissues after the fat has become quite cold.

For the examination of adipose tissue the mesentery of a small animal, as a mouse, rat, or rabbit, is most satisfactory. A small portion of the structure, with the smallest particle of fat attached, should be cut off with fine scissors, placed in a little water on the glass slide, and at once gently covered with the thin glass, without the least pressure. The low power should be used first, to obtain a general idea of the arrangement of the structure, and afterwards, the higher one for the purpose of tracing the details.

The drawing represents an average specimen, such as the student will find no difficulty in preparing.

Fig. 40.

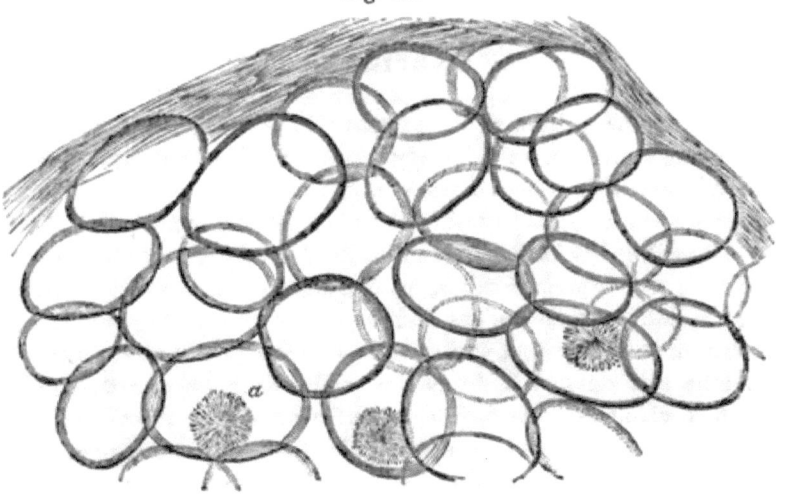

Adipose tissue. *a*, Starlike appearance from crystallisation of fatty acids.—High power.

Very good specimens of fatty tissue may be obtained by cutting off a small piece of subcutaneous cellular tissue of the calf, placing it in a drop of water on the slide, slightly teazing it out, and then gently applying the covering glass in the usual manner.

Fine sections may also be made from the fat of cold roast beef by cutting it with a razor, selecting the soft surface fat in preference to the suet, which is not good for the purpose.

Isolated fat cells are best obtained from marrow, a minute portion of which should be placed in a drop of water on the slide, and teazed out carefully so as not to injure the delicate cells; the thin glass cover must be very lightly applied, and the specimen examined with a high power.

If too much pressure has been used, the vesicles will be ruptured, and then only masses of oil and numerous granules will be visible.

CARTILAGE.

Cartilage or gristle, as it is commonly termed, is a dense non-vascular structure, of a pearly white or slightly yellow colour, distributed with tolerable uniformity throughout the body wherever a certain solidity combined with elasticity is necessary.

Thus, it exists in the anterior walls of the thorax, in the trachea, larynx, and bronchial tubes, in the nose, ears, eustachian tubes, and eyelids. It covers the ends of bones where they are in apposition to form joints, and thus lessens the effects of concussion, and in the embryo it forms the entire skeleton, a considerable part of which remains cartilaginous for some time after birth.

Cartilage is covered by a dense fibrous membrane termed perichondrium, excepting that kind known as articular cartilage, which receives a layer of epithelium from the synovial membrane extending over the whole surface of the joint in the fœtus, but in the adult confined to the circumferent margin, in consequence of the central wear which takes place as soon as the parts are subjected to friction during the movements of the animal's limbs.

Cartilage is distinguished as *permanent* when it continues unchanged, either covering the ends of bones (articular cartilage), or forming the walls of cavities (membraniform cartilage); and *temporary* when it is subject to the process of ossification, during the development of the animal textures.

Besides the ordinary kind of cartilage temporary and permanent, there are two modifications of the tissue existing in certain parts of the body, viz. cellular cartilage, which is composed of cells lying close together in a mesh formed of fine fibres; and fibro-cartilage, in which the cartilage cells are distributed in a matrix everywhere pervaded by fibrous tissue, either of the white or yellow variety.

For the microscopic examination of ordinary cartilage the head of the humerus of an adult animal should be taken; that from an ox may be easily obtained from the butcher. Fine sections should be made with a razor near the surface and parallel to it; the specimen must next be transferred to a drop of water in the centre of the glass slide, covered in the usual way with a piece of thin glass, and viewed first with a low, and afterwards with a high power.

The illustration will convey a general idea of what the observer should see, but it is necessary to notice that the form and arrangement of the cells will be modified by the direction and depth of the section.

Fig. 41.

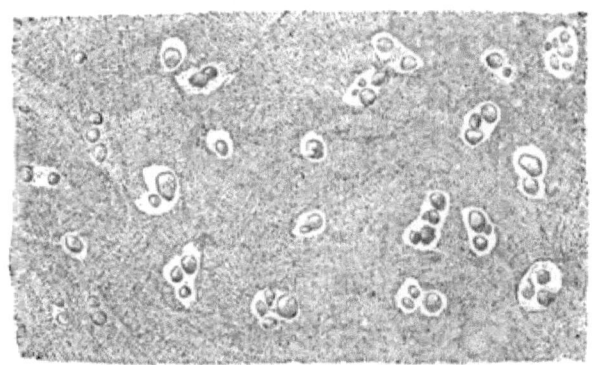

Cartilage of humerus of ox.—Magnified 100 times.

Examined with a low power, ordinary cartilage looks like a homogeneous membrane studded over with numerous round, oval, oblong, semilunar, and irregularly shaped corpuscles; under a high power the membrane or matrix loses its homogeneous character, and appears to be either finely granular or faintly striated.

In articular cartilage the matrix is granular, the cells or corpuscles are collected in variously-sized groups of two to six, lying in little hollows, or lacunæ, which the cells occasionally do not completely fill. A membrane is said to line the lacunæ, and to be rendered visible by boiling the cartilage for some hours, and afterwards treating it with acetic acid. This view makes the cartilage cell homologous with the primordial utricle.

The cells sometimes contain granular matter, and occasionally fat-globules. They also possess distinct nuclei, varying from $\frac{1}{4000}$ to $\frac{1}{2400}$ of an inch in diameter, and sometimes also nucleoli. Parent cells will be observed, enclosing several young ones, as the cells multiply by the endogenous process, already explained.

In sections taken near the surface of the joint, the groups of corpuscles are flattened and run parallel, but in deeper sections, that is to say, nearer the bone, the groups are oblong and narrow, sometimes appearing like strings of beads.

The variety of permanent cartilage attached to the ribs differs from articular cartilage in several particulars. In the first place, the corpuscles are very much larger, have better defined nuclei, and are more regularly grouped together. In the second place, rib cartilage has a tendency to become fibrous, more particularly as the subject advances in years.

These peculiarities are beautifully shown in the illustration, taken from a specimen of rib cartilage from an aged subject.

Fig. 42.

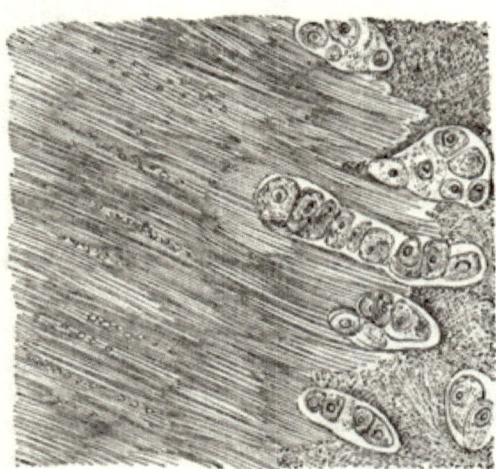

Cartilage of rib from a man seventy-six years old, showing the development of fibrous structure in the matrix. In several portions of the specimen two or three generations of cells are seen inclosed in a parent cell wall.—High power.

Temporary cartilage may be readily distinguished from permanent, by the absence of the *groups* of corpuscles, the cells being distributed at nearly equal distances apart in the intercellular substance. The cells are mostly round or oval, and the nuclei are granular.

Specimens for examination may be obtained from any very young animal in whom the ossific process is not complete.

CELLULAR CARTILAGE.

The humerus or femur of a calf will furnish good examples. The peculiarities of temporary cartilage, however, are best studied during the investigation of the process of ossification, which will shortly be considered.

Cellular Cartilage.—This variety of cartilaginous structure is found in the ears of all small animals from the size of the rat downwards, and even in the tips of the ears of larger animals, such as the rabbit and dog.

In order to examine cellular cartilage, the ear of a mouse should be taken, and first dried until sufficiently firm to be cut with a razor; fine slices may then be shaved off transversely, and placed in a drop of water on the glass slide, covered with the piece of thin glass in the usual way, and examined with a high power.

The drawing indicates the appearance which the specimen should present if properly prepared.

Fig 43.

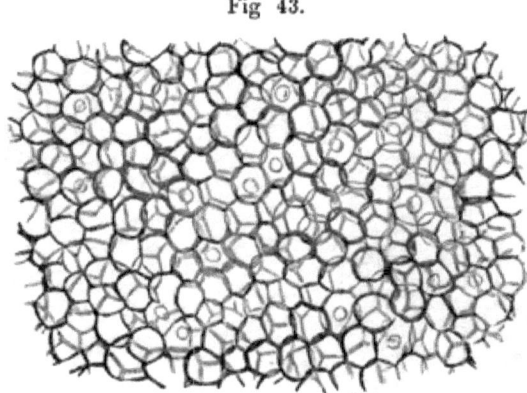

Cartilage from mouse-ear.—Magnified 100 times.

The cells are large, and appear to be packed closely together in cavities in the matrix, but, when minutely examined, it will be found that the corpuscles do not impinge upon each other, but are surrounded by clearly defined fibres, so arranged as to form a very beautiful and perfect network, in the meshes of which the cells are placed.

In many parts of the specimen empty spaces will be noticed,

from out of which the cells have fallen. Generally the drying process causes the cell contents to shrink, consequently a nucleus is seldom visible.

Yellow Fibro-Cartilage.—This form of fibro-cartilage is found in the ears of large animals, in the epiglottis, the cornicula laryngea, and in the eyelids. It is much more opaque and more flexible than other kinds of cartilage, and has little tendency to ossify.

For examination, the ear of a large animal, such as a horse or ox, should be taken and slightly dried, after which, from the root portion, fine sections may be made with a razor. One of the best sections should be placed in a drop of water on the glass slide, covered with the thin glass, and examined, first with the low and afterwards with a high power. In a good specimen the appearances presented in the drawing will be recognised.

Fig. 44.

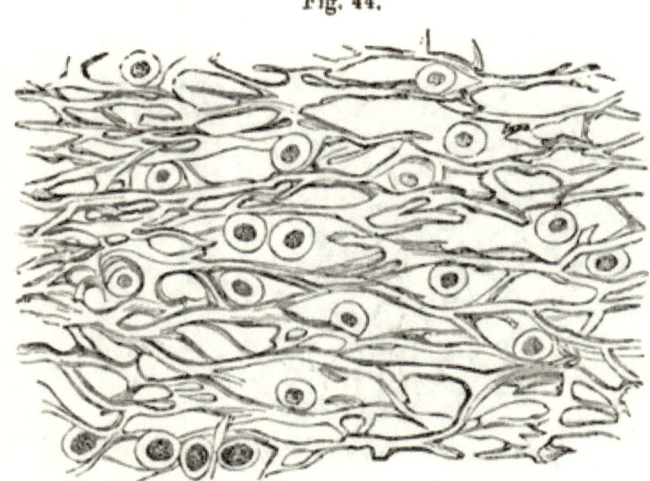

Yellow cartilage (ear of horse).—High power.

The cells are very numerous, and are distributed in a matrix or network of large fibres, which diverge from the centre to the circumference.

The fibres look like yellow elastic fibrous tissue, possess the same double outline, anastomose and branch in all directions, and gradually become finer as they approach the surface. They are not acted on by acetic acid.

These yellow elastic fibres give to the cartilage its yellow colour and remarkable flexibility.

White Fibro-Cartilage.—Another kind of fibro-cartilage is constituted of white fibrous tissue, with cartilage cells. It occurs between the bodies of the vertebræ as a connecting medium. As interarticular discs it is found between joint surfaces, and it sometimes surrounds and deepens a joint-cavity, being attached to its margin, as in the case of the acetabulum.

A specimen may be obtained by cutting fine sections with a razor from the centre portion of an interarticular disc from the patella joint of any animal, ox or sheep for example. The section must be placed in a drop of water on the slide, covered in the ordinary way, and examined with a high power.

The oval or circular cartilage corpuscles will be seen surrounded by an abundance of white fibrous tissue, as shown in the illustration.

Fig. 45.

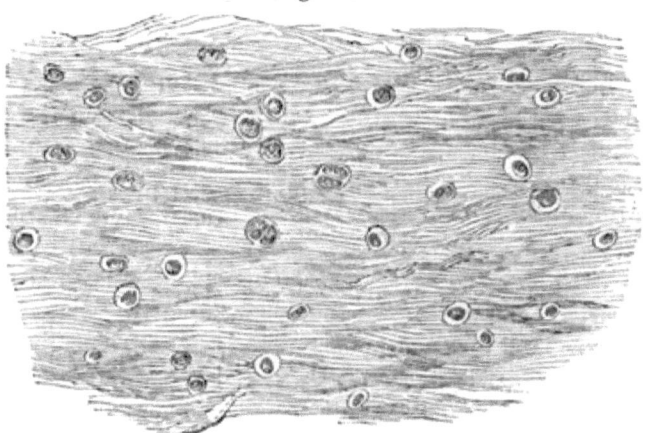

White fibrous cartilage from the semilunar disc of the patella joint of an ox.—Magnified 100 times.

BONE.

Bone is hard white unyielding structure which in the vertebrata forms the basis of the animal frame, supporting or surrounding the softer textures.

Although varying much in external form in different parts of the skeleton, it has everywhere a perfectly uniform composition, chemical and physical. Bone consists of earthy and animal matters so intimately combined, that the removal of either does not destroy the form of the bone, if the process of separation has been carefully conducted.

Thus the animal matter may be removed by calcination in an open fire, leaving nothing but the phosphate and carbonate of lime; and notwithstanding the bone, by this process, loses its tenacity and becomes exceedingly brittle, the form, in its minutest particulars, is preserved.

The earthy constituents, on the other hand, may be all dissolved out by dilute hydrochloric acid, which leaves only the animal matter of the bone, without altering its form, quite perfect. Its hardness, however, is destroyed, and the tissue becomes as soft and elastic as a piece of cartilage. If a long bone, as a rib, be so treated, it may be tied in a knot with ease.

On becoming dry, the macerated bone shrivels, and assumes the density of horn.

The interior of a bone is of a spongy or cancellated structure, particularly at the ends. The outer portion of the bone is much more dense than the internal part, and is described as the cortex or shell of the bone.

Microscopic Examination of Bone.—The study of bone should commence with sections of the softened structure. For this purpose a piece of bone should be placed in a mixture of hydrochloric acid one part, to water sixteen parts, and allowed to remain for a week or two, after which it must be thoroughly washed in clean water to remove the acid. Sections may now be made in any direction with a sharp knife or razor; the same manipulation will be necessary as for other tissues. The sections must be transferred to a drop of water on the slide, covered in the usual way, and examined first with a low, and afterwards with a high power.

A transverse section from a long bone, prepared in the manner described, is shown in the drawing, which very clearly indicates the lamellated structure of bone.

Fig. 46.

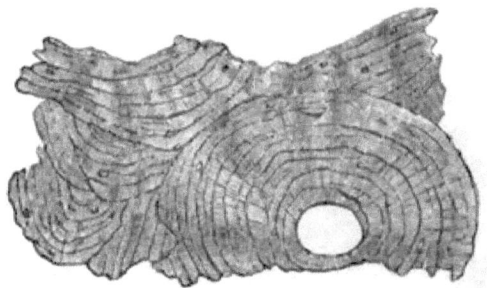

The laminæ as they appear after the removal of the earthy matter by the action of acid.

Two series of lamellæ may be demonstrated in bone after maceration in dilute acid; a larger system surrounding the medullary canal, and a smaller system surrounding the Haversian canals, both of which may often be seen in the same section.

In macerated bones the lamellæ of the larger concentric system may be peeled off in layers, which are pierced with fine apertures; these are caused by the canaliculi, to be afterwards noticed. In some parts larger apertures are seen, through which fibres or bunches of fine fibres pass, pinning the several layers of bone together. These are the perforating fibres (Sharpey).

The outermost of these layers, being near the periosteum, are termed periosteal layers: the innermost ones, being close to the canal containing the marrow, are called medullary layers.

Sections of the compact tissue of fresh bone may be obtained ready prepared, or they may be made for examination by cutting thin slices with a fine saw, then glueing them on to a piece of level wood, and reducing them to the requisite thinness, first with a file and then on a hone with water, next smoothing the surface with fine emery paper, and lastly polishing with a piece of leather dipped in putty powder. The specimen must afterwards be removed from the wood by the aid of a little warm water, and refastened on the polished surface, in order that the unfinished side may be treated in the same way; after which it may be again unglued, washed carefully with warm water by means of a small brush, and allowed to dry. It may be examined in a dry state on the slide, covered with a piece of thin glass; or it may be placed in a drop of water which will render it more transparent.

The illustration represents a portion of a long bone, and shows what the observer should see in a properly prepared specimen.

Fig. 47.

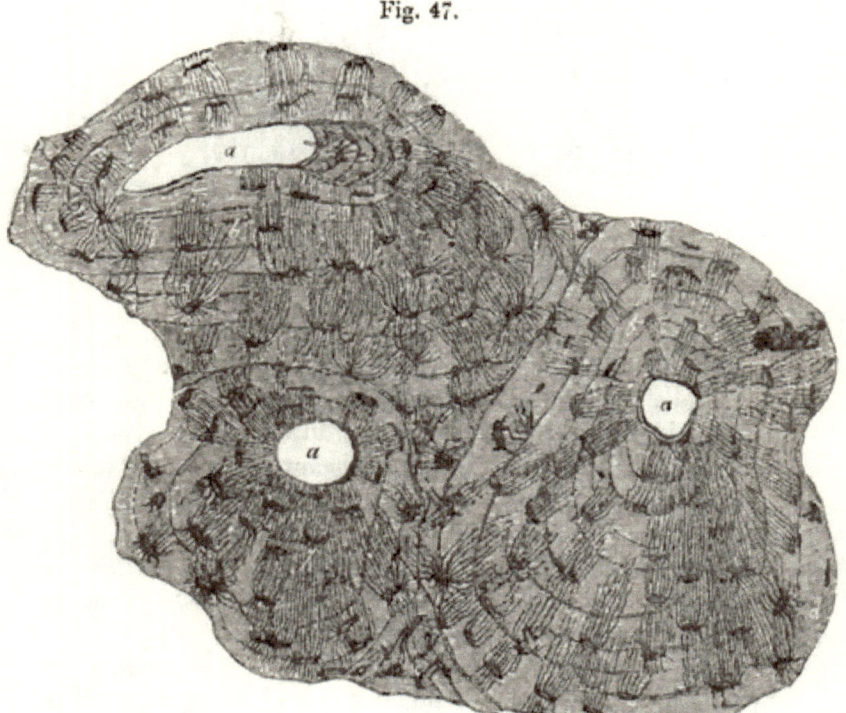

Transverse section from the dense portion of the femur. *a.* Haversian canal. *b.* Concentric laminæ. *c.* Laminæ of connection. *d.* Corpuscles, with their system of tubes. The parts marked *a*, *b*, and *d* constitute an Haversian system. The figure includes three systems with laminæ of connection uniting them.

The Haversian canals will be observed in all parts of the tissue varying in size from $\frac{1}{2000}$ to $\frac{1}{200}$ of an inch in diameter, the largest being nearest to the medullary canal. In shape, they are round, oval, or oblong, according to the line of section.

Each canal is surrounded by rings, none of which are complete; never, in fact, extending quite round the canal, but running one into the other at various parts.

Under a high power there will be observed numerous irregularly shaped cavities or hollows termed lacunæ, with fine

Fig. 48.

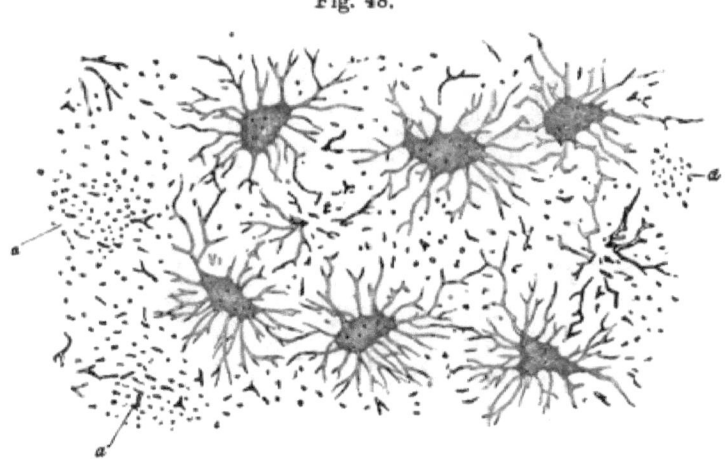

Section from the parietal bone. Lacunæ with the canaliculi seen from the surface; magnified 450 times. The points upon the cavities, or between them, belong to the cut canaliculi, or are the openings of the latter into the lacunæ. *a, a.* Groups of transversely cut canaliculi, each belonging to a lacuna which has been destroyed in the preparation of the section.

canaliculi which radiate from them in all directions to terminate in the Haversian canals.

Those of the lacunæ, which are close to the medullary canal, have their canaliculi opening into that cavity.

Virchow considers that each lacuna is occupied by a ramified nucleated cell (analogous to the corpuscles of tendon and muscle), the walls of which line the cavity while the branches are prolonged through the canaliculi.

By means of the complete and intricate distribution of the canals of the Haversian system, fluids can pass into the most compact parts of the osseous tissue.

A longitudinal section, as represented in the next drawing, will show the course of the Haversian canals to be in the direction of the long axis of the bone, or obliquely to it.

Fig. 49.

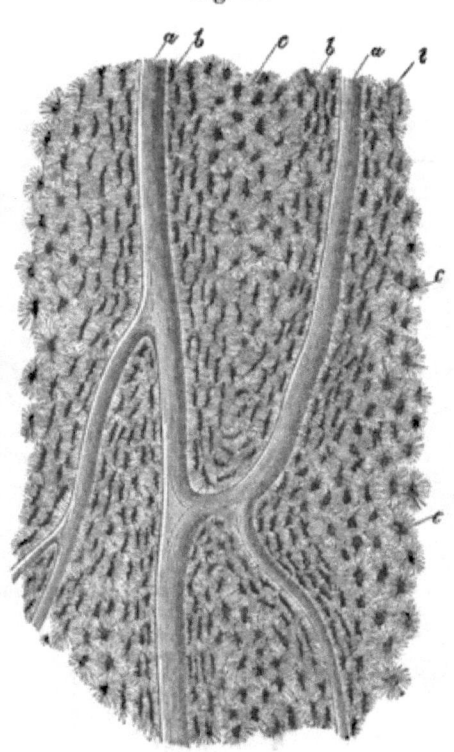

Section parallel to the surface, from the diaphysis of a human femur; magnified 100 times. *a.* Haversian canals. *b.* Lacunæ seen edgeways, belonging to the lamellæ of the same. *c.* Others seen flatways, in lamellæ which are cut parallel to their surface.

The earthy matter of bone may be seen by placing a minute portion of calcined bone in a drop of water on the slide, and reducing it to powder by rubbing it with another piece of glass. A drop of diluted hydrochloric acid should then be added to dissolve the broken granules, and in a few minutes the dilute acid is to be removed carefully with blotting-paper, and a little water added in its stead; the specimen is then to be covered and examined with a high power.

The earthy matter will appear as shown in the drawing.

Fig. 50.

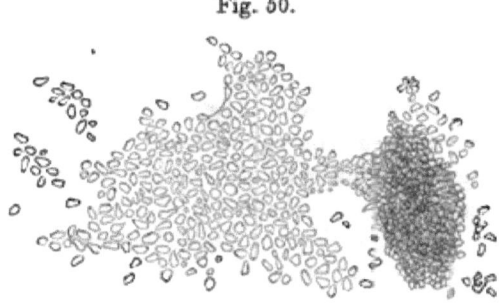

Earthy matter obtained by depriving the bone of its animal matter.

Development of Bone.

Ossification of Cartilage.—The rudiments of the osseous system are developed very early among the embryonic structures; the base of the skull, and the vertebræ being the first to appear. These, like other parts of the embryo, are composed of a blastema, in which elementary cells are formed. The cells are converted into cartilaginous structure, which ultimately becomes bone, by a systematic deposit of earthy matter.

Bone does not always commence from cartilage; the cranial bones, for example, furnish an example of ossific development going on in a membrane quite distinct from cartilage.

Dr. Sharpey remarks that even in long bones in which ossification undoubtedly commences, and to a certain extent proceeds in cartilage, there is much less of the increment of the bone really owing to that mode of ossification than is generally believed.

Two separate modes of ossification are therefore to be distinguished : the intra-membranous and the intra-cartilaginous.

M. Ollier of Lyons, following up earlier observations by Mr. Syme, related, some time since, a series of experiments, and exhibited specimens tending to show that the periosteum is the true secreting membrane of bone.

'In one specimen the radius had been wholly removed, leaving the periosteum entire, and the bone was reproduced. Portions of periosteum were transplanted round the muscles of the thigh, and they secreted bone. This was done in the form of strips twisted round, and they secreted bone of a spiral shape. In fact, any form of bony secretion can be obtained, almost at the fancy of the operator.'

M. Ollier transplanted entire bones and portions of bones with their proper membranes, and they retained their vitality; but if the membranes were removed in these experiments, the bones died, and suppuration ensued. All these osseous secretions possessed the true and normal character of bone.

Ossification begins in the middle of the bone and proceeds to the extremities, which remain for some time cartilaginous. Separate ossification at length commences in the ends of the bone forming epiphyses, which are ultimately united to the body of the bone.

Newly-formed osseous tissue is red, and highly vascular, the vessels extend a short distance into the adjoining cartilage, ramifying in canals much larger than the vessels which they contain.

Examination of Ossifying Cartilage.—Take the humerus or femur of a calf, and make very thin sections with a razor in a longitudinal direction through a portion of the head of the bone, select a piece which includes both bone and cartilage, place it in a drop of water on the slide, cover it with a piece of thin glass, and examine it, first with the low and afterwards with the high power.

Great care is required in making the sections, and it may be necessary to examine several before a satisfactory observation can be made.

In a good specimen the structures shown in fig. 51 will readily be distinguished.

Fig. 51.

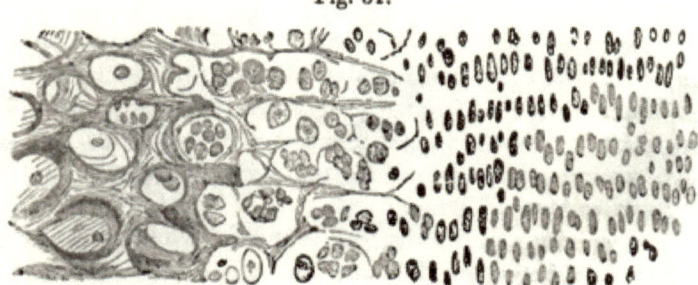

Longitudinal section of ossifying cartilage from the humerus of a calf.

At a distance from the point of ossification the cartilage cells will be seen to be uniformly disseminated in the matrix, but as they approach the ossifying portion, they are collected into oblong groups, the cells having their long diameter transversely; in the immediate vicinity of the bone they become greatly enlarged and more rounded.

Into the clear spaces, between the groups of cells, dark lines of bone shoot up. The bony tissue, as it advances into the cartilage, forms short tubular cavities (areolæ).

Earthy matter is first deposited in the matrix, and surrounds the cartilage cells, as may be seen by making a transverse section of the ossifying cartilage, represented in the illustration.

Fig. 52.

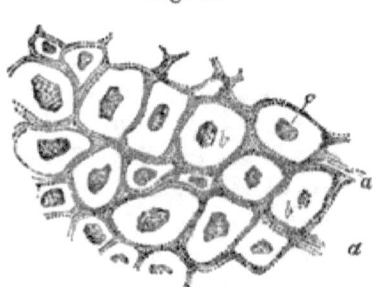

Transverse section of temporary cartilage in the first stage of ossification. *a, a.* Intercellular tissue ossified. *b.* The transparent parietes of the enlarged corpuscle. *c.* The central nucleus, which in the specimen from which this figure was taken, was granular.

As the ossific process advances the bony areolæ open into each other by partial absorption of their walls, and larger cavities then result, which become filled with bloodvessels and granular cells, probably derived from the cartilage cells, the corpuscles of which have been opened. The original cartilaginous matter disappears by absorption, and the walls of the cavities become thickened by layers of secondary bone, deposited by the granular cells. This lamellar deposit is thin when the cavities remain to form cancellated structure, but in other parts it may nearly fill up the space with consecutive layers, leaving a vascular canal in the centre, and in this way Haversian systems may be produced, although most of these are formed independently of preceding cartilage, by subperiosteal ossification, which goes on at the same time and forms the compact tissue of the shaft.

MUSCLE.

MUSCULAR STRUCTURE is distinguished by the property of contractility, which is excited by certain stimuli.

Two kinds of muscle exist in the animal body, differing from each other in structure and function.

1. Voluntary muscle, which contracts under the influence of the *will*.

2. Non-voluntary muscle, which acts independently of the *will*.

Examination of Striated or Voluntary Muscle.—For the study of this structure, a portion of boiled beef or horseflesh is to be obtained, and systematically dissected, in order that each of the three parts into which voluntary muscle is divisible may be separately examined, viz.: 1. Fasciculi, which are bundles of fibres; 2. Fibres, which are bundles of fibrillæ; 3. Fibrillæ, which appear like rows of fine particles, called 'sarcous elements.'

Fasciculi are visible to the unaided eye, constituting the fibres seen in boiled beef. The coarse or fine texture of muscle depends upon the size of the fasciculi, which vary in diameter according to the number of fibres united to form them.

By the aid of a common lens the prismatic form of the fasciculi, their variation in size, and general arrangement, may be distinctly seen.

Fig. 53.

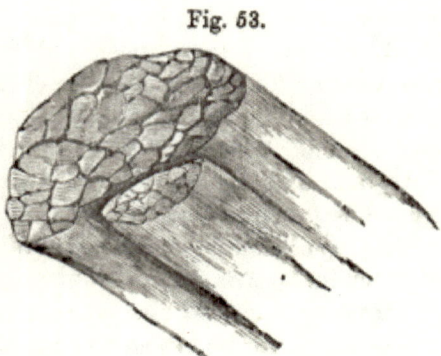

Fasciculi of Muscle.—Magnified 10 diameters.

Each fasciculus is invested by a sheath of areolar tissue, termed 'Perimysium,' which may be seen in transverse sections, made by drying a piece of half-boiled horseflesh, about three inches long by one inch in diameter, until it will bear the pressure of the razor. Fine sections may then be cut and placed on the slide in a little water; in a short time the natural condition of the tissues is in a great measure restored, and the object may be covered, and viewed with a low power.

If the section is sufficiently thin, the fasciculi, with the investing membrane and the ends of the component fibres, will be very distinct.

Fig. 54.

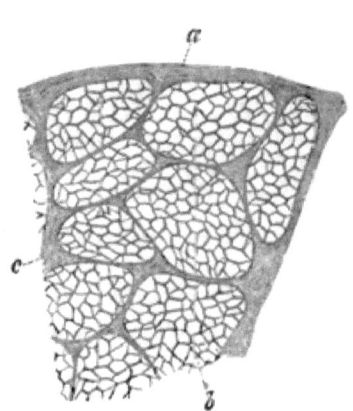

Transverse section from the sterno-mastoid in man; 50 times magnified. *a*. External perimysium. *b*. Fasciculi. *c*. Internal perimysium.

Generally the fasciculi run straight from one end of the muscle to the other, excepting in those instances where the tendon extends some distance into the substance of the muscle, in which case, numerous *short* fasciculi are attached to one or both sides of the tendinous cord, like the fibres of a feather to its stem, giving rise to the terms penniform, semi-penniform, and compound penniform, according to the arrangement.

Examination of the Fibres.—Take a piece of boiled horseflesh (ordinary cat's meat) strip off a small portion of one of the fasciculi, place it in a drop of water on the slide, and teaze it well out, seeking to separate the fibres lengthwise, apply the covering glass, and examine under the low power.

The object will resemble a bundle of sticks, with here and there a single one.

Under the high power the fibres, with their transverse striæ, will be distinguished, and in some parts will be seen to be split up at their ends into fibrillæ.

Fig. 55.

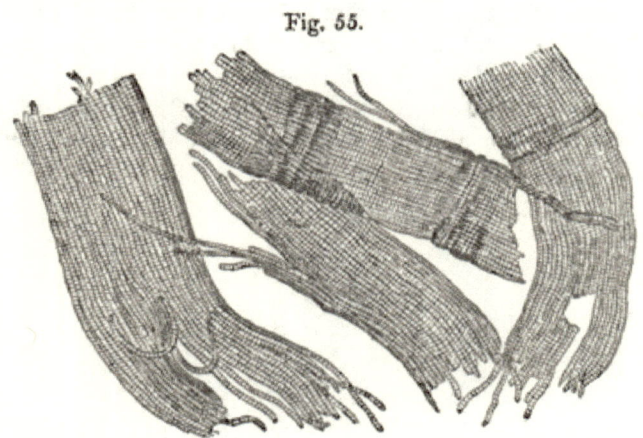

Muscular fibres of Horse.—Magnified 200 diameters.

In size the fibres vary but little, the average being $\frac{1}{400}$ inch. Mr. Bowman finds the size, in the male, to be $\frac{1}{362}$ inch, and in the female, $\frac{1}{454}$.

During the examination of the specimens of voluntary fibres, particularly from cooked flesh, a peculiar cleavage into discs will often be observed, as shown in the drawing (fig. 56).

The appearance was recently noticed in beef that had been preserved for some months, nearly all the fibres showing the tendency to this transverse cleavage.

The dark lines, or striæ, are $\frac{1}{9400}$ inch apart, and are particularly characteristic of voluntary muscle; but they nevertheless also exist in the muscular fibres of the heart, in the pharynx, and in the upper part of the œsophagus.

Striated fibres do not anastomose nor divide, excepting in the heart, in the tongue of the frog, and in the facial muscles of some animals.

Each fibre is invested by a transparent structureless membrane or sarcolemma, corresponding to the internal perimysium.

Fig. 56.

A. A muscular fibre, or primitive fasciculus, breaking up in the transverse direction, into discs, 350 times magnified. It exhibits distinct transverse and fainter longitudinal striæ. The discs, of which B represents one more magnified, are granulated, and consist of the sarcous elements of Bowman, or, according to other authors, of small pieces of the fibrils. After Bowman.

In fishes the membrane, being tough and elastic, may easily be demonstrated, as it is often left entire when the substance of the fibre is torn across by being stretched.

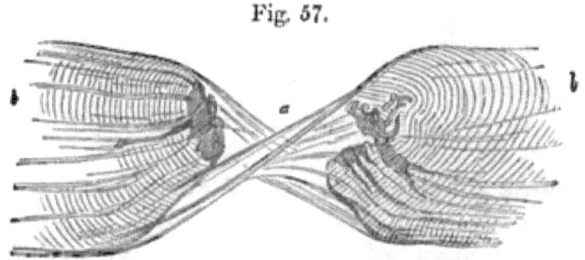

Fig. 57.

Fragments of an elementary fibre of the Skate, held together by the untorn but twisted sarcolemma. *a*. Sarcolemma. *b, b*. Opposite fragments of the fibre.

A drop of acetic acid will render the nuclei of the sarcolemma apparent. In the fœtus they can be seen without the aid of chemical agents.

Besides corpuscles like cells or nuclei lodged among the fibrils within the fibre, others exist between the fibres ('fusiform corpuscles,' Ellis), which correspond in aspect, and probably in function, with the corpuscles of connective tissue.

In the muscles of the frog, Welcher discovered peculiar corpuscular bodies between the fibres, as depicted below.

Fig. 58.

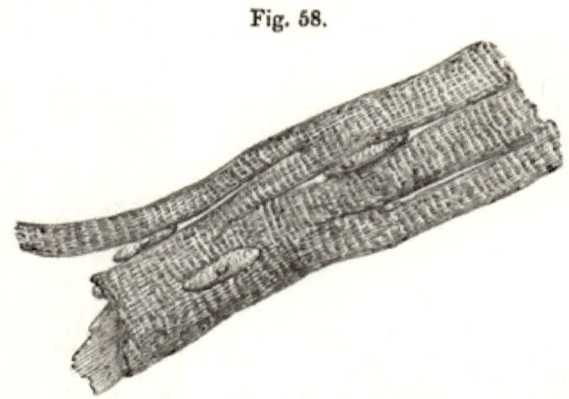

Portion of muscle of a frog, showing nucleated cells between the fibres.

Examination of Fibrillæ.—Among mammalia the pig furnishes the best examples of fibrillæ, but their isolation is extremely difficult. They are more easily prepared from the insect. A good plan is to cut open the thorax of a fly, and remove a portion of the large yellow muscle which lines its walls; this should be teazed out in a drop of water on the slide, and covered with the thin glass as usual.

A power of 400 to 800 diameters will be required to render the elements distinctly visible.

Each fibril consists of a row of minute particles (sarcous elements) strung together like beads (fig. 59).

When highly magnified these particles appear as dark squares with bright spaces between them; sometimes faint lines may be seen crossing the bright spaces. Under a still higher power the dark squares are seen to be somewhat constricted in the middle and surrounded by a bright area.

The dark aspect of the central part probably arises from its refracting the light differently from the surrounding portions, as

Fig. 59.

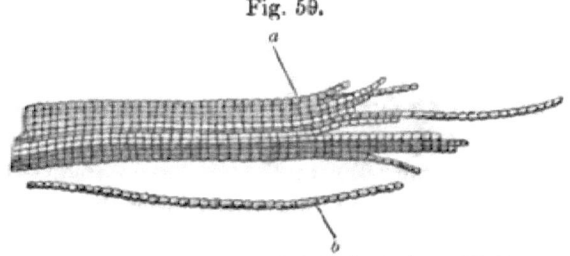

Primitive fibrils from a fibre, or primitive fasciculus; 600 times magnified. *a.* A small bundle. *b.* An isolated fibril.

by altering the focus the dark part becomes light and the light dark.

Bloodvessels are exceedingly numerous in voluntary muscles, and form a fine network among the fibres. None of the branches penetrate the sarcolemma.

In the illustration the arrangement is very correctly shown.

Fig. 60.

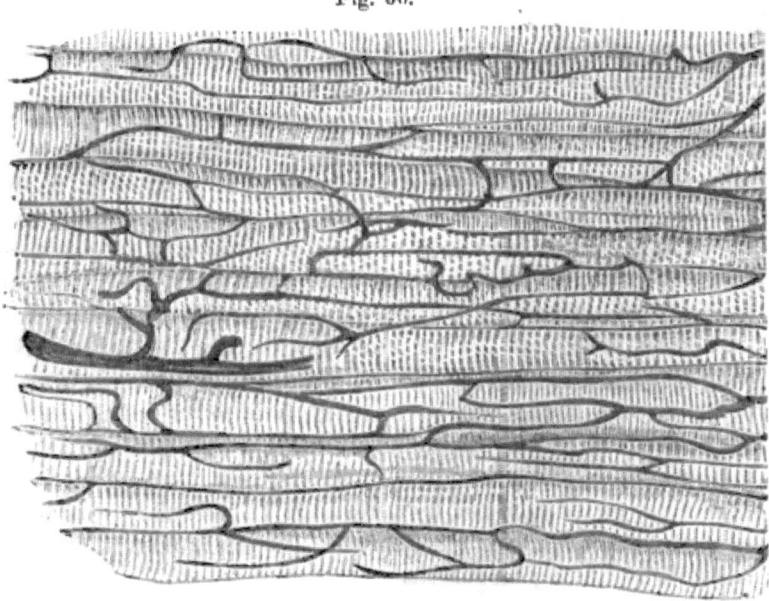

Vessels of muscular tissue, from an injected preparation.

Nerve fibres are said to be distributed to the elementary fibres of striated muscle. In connection with the minute nerve twigs little oval bodies are found, and are considered to be the means of communication between the nerves and fibres of the muscle.

Involuntary Muscle, Striated Variety.—A portion of a heart should be boiled for a few minutes in acetic acid, and afterwards dissected and carefully teazed out in the manner directed for the preparation of voluntary fibres; the transverse striæ will readily be distinguished; even in an unprepared heart there is ordinarily no difficulty in seeing them.

In the sheep's heart branching fibres may often be observed.

Dr. Harley has distinguished smooth muscular fibres in the heart of the bird.

The fibres of the heart in the early embryo are non-striated, being composed of elongated cells with nuclei, which are rendered very distinct by a drop of acetic acid.

Non-striated Variety.—Non-voluntary muscle exists in all parts where movements occur independently of the will; in the walls of the intestinal tube; in the trachea and bronchia; in the bladder and ureters; in the uterus and corpora cavernosa; in the ciliary muscle and in the iris; in the middle coat of arteries; and in parts of the skin.

The fibres of involuntary muscle are united to form fasciculi of various sizes, which cross and interlace in all directions, the fibres only running in a straight course and parallel to each other. For the purpose of examination take a piece of intestine, and macerate it for two days in a mixture of nitric acid, one part to four parts of water. This causes the tissues to swell and become yellow, while it renders the separation of the fibres more easy.

A small portion of the specimen should then be cut from the edge by means of curved scissors, and teazed well out in a little water on the slide; the covering glass is to be applied in the usual way, and the object examined under a high power.

Non-striated fibres are pale in colour, round or prismatic in shape, but easily flattened by pressure. In size they vary from $\frac{1}{7000}$ to $\frac{1}{3500}$ of an inch in diameter. They are marked at short intervals by oblong corpuscles.

If the preparation has been properly made, a number of pale fibres with nuclei will appear, as in the drawing, which is taken from a preparation of the muscular coat of the human colon treated with acetic acid.

INVOLUNTARY MUSCLE.

Fig. 61.

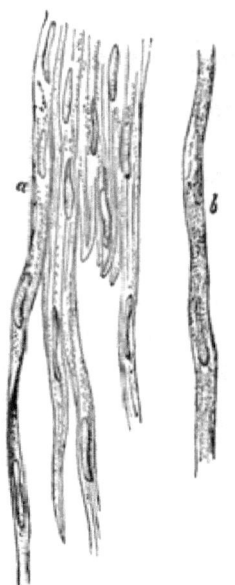

Non-striated elementary fibres from the human colon. *a.* Treated with acetic acid, and showing the corpuscles. *b.* Fragment of a detached fibre, not touched with acid.

The substance of the fibre is translucent or finely granular, the granules being arranged in longitudinal lines. Occasionally faint striæ are seen, particularly in the fibres of the fœtal heart.

Some authorities consider involuntary fibres to be constituted of elongated nucleated cells cemented together.

In examining the fœtal heart, the separation of fibres into elongated cells is very distinctly seen under a moderately high power.

THE INTEGUMENT.

The skin consists of connective tissue, with numerous nerves and vessels, and an external covering of cells.

By means of sections, which are easily made, the student may examine the constituents of the skin in their relative position, and subsequently, by scraping from the surface, and by teazing out portions of the deep-seated structures, he will be enabled to investigate the elements separately.

In order to make sections, a portion of the integument of any animal must be slowly dried until sufficiently firm to bear the pressure of the razor. When properly prepared it may be cut as readily as a piece of soft horn.

Thin slices should be cut with the razor, and transferred to a drop of water on the slide. In a few seconds the section will be seen to imbibe the fluid and become spread out flat upon the glass, having regained, in great part, its natural appearance. The thin covering glass must next be applied in the usual way, and the specimen examined, first with a low power.

If the section has been properly prepared the several parts will be apparent, as shown in the illustration (fig. 62).

1. The epidermis, or cuticle, or scarf skin.
2. The dermis or cutis vera, or true skin.
3. Hairs, with their follicles, glands, and ducts, more or less perfect, according as the preparation is successful.

The epidermis is entirely composed of epithelial scales.

It is insensible and non-vascular, containing neither nerves nor vessels; the only structural objects visible in it, besides the hairs and their follicles, are the tubes of the perspiratory and sebaceous glands, and even these are seldom seen until the specimen has been specially prepared with oil of turpentine.

The palms of the hand and the soles of the feet are covered with the thickest layers; sometimes in these situations it is horny as well as thick. The cells forming the cuticle are arranged in irregular rows.

In the deepest layers the cells are oblong or prismatic and set vertically, the succeeding ones are round, and, like the deeper

Fig. 62.

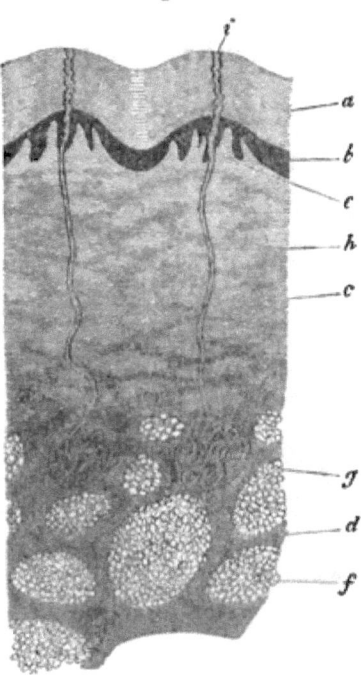

Vertical section through the skin of the ball of the thumb, transversely through two ridges of the cutis; magnified 20 times. *a.* Horny layer of epidermis. *b.* Its mucous layer. *c.* Corium. *d.* Panniculus adiposus (upper parts). *e.* Papillæ of the cutis. *f.* Fat globules. *g.* Sudoriferous glands. *h.* Their canals. *i.* Sweat-pores.

cells, dark in colour, containing granular matter, with well-marked nuclei. As they approach the surface they become flattened, lighter in colour, from the removal of their granular contents, hard, dry, and insoluble in acetic acid.

To render the cellular character of the epidermis more apparent, a drop of liquor potassæ should be allowed to flow under the covering glass and act slowly on the skin. In a short time the epidermis will be seen to swell, and the cells gradually separate from each other, permitting their characters to be clearly made out. Layer upon layer will be observed to cover the papillæ of the true skin, following their undulations with unvarying regularity.

The pigment cells in the skin of the negro are principally found in the rete mucosum, so that when a portion of this skin is examined under the microscope, it appears as if there were a continuous dark band (seen in the illustration) following the undulations of the papillæ upon which the cuticle is accurately moulded.

Fig. 63.

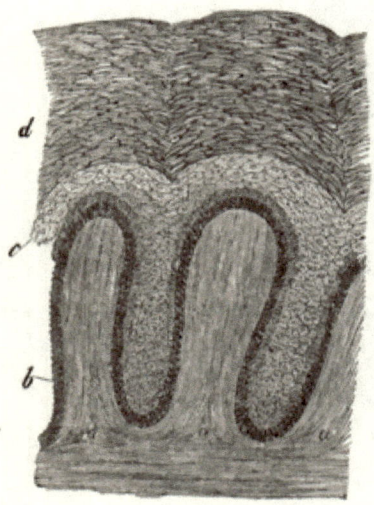

Perpendicular section through the skin of a negro (from the leg). *a.* Papillæ. *b.* Deepest intensely coloured layer of elongated cells of the rete mucosum. *c.* Upper layer of the rete mucosum. *d.* Horny layer. —Magnified 150 diameters.

Dermis, Cutis Vera, or True Skin.—The true skin is divided into the following parts :—

Cutis Vera { Corium { Papillary layer. Reticular layer. } Tela cellulosa subcutanea. Panniculus adiposus. }

The true skin is vascular and sensient, and attached to the subjacent parts by areolar tissue.

Immediately below the epidermis is the basement membrane, called '*Membrana propria*,' a delicate structureless film, only perceptible in the embryo. Under the basement membrane is the papillated layer of the corium, a dense firm tissue, containing the ultimate expansion of vessels and nerves, and the upper parts of the hair follicles, and tubes of cutaneous glands.

To study the papillary part of the corium it is necessary to remove the epidermis by boiling or maceration, or, when the portion of skin is very thin, by the action of acetic acid or caustic soda.

Sections of the corium must then be made by means of the double knife, floated on to the slide in the manner previously directed, and covered in the usual way.

Under the low power the papillæ will be readily seen, varying somewhat in form in different parts of the integument, but presenting the general characters shown in the drawing.

Fig. 64.

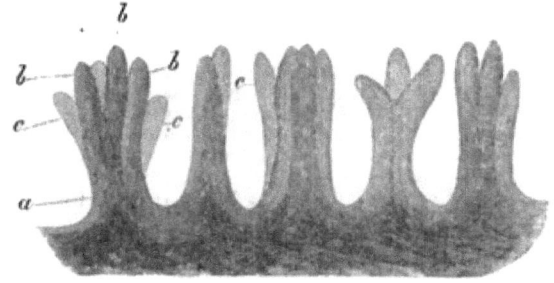

Compound papillæ of the surface of the hand, with two, three, and four points. *a.* Base of a papilla. *b b.* Their separate processes. *c c.* Processes of papillæ, whose base is not visible. —Magnified 60 diameters.

A more minute examination of the structure of the papillæ will necessitate the isolation of a single one by means of dissecting needles.

The specimen must be carefully transferred to a drop of water on the slide, covered with a piece of thin glass, and examined with the high power. A drop of acetic acid allowed to flow under the covering glass, and act upon the structure, will render the nuclei of the outer layer distinct, as seen in the illustration (fig. 65).

In the interior of a few of the papillæ there will occasionally be observed a solid body of an oval shape, which has received the name of tactile corpuscle. It is full of nuclei as will be seen.

Round the corpuscle a nerve is sometimes seen to be winding previous to entering its structure.

HISTOLOGY.

Fig. 65.

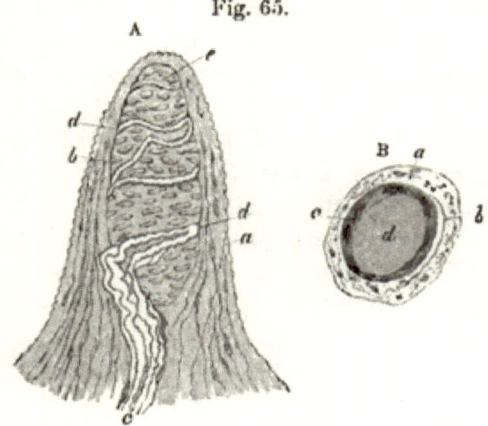

A. Side view of a papilla of the hand. *a.* Cortical layer, with plasm-cells and fine elastic fibres. *b.* Tactile corpuscle, with transverse nuclei. *c.* Small nerve of the papilla, with neurilemma. *d.* Its two nervous fibres running with spiral coils around the tactile corpuscule. *e.* Apparent termination of one of these fibres. B. A tactile papilla seen from above so as to show its transverse section. *a.* Cortical layer, with plasm-cells. *b.* A nerve-fibre. *c.* Outer layer of the tactile body, with nuclei. *d.* Clear interior substance. From the human subject; treated with acetic acid.—Magnified 350 times.

The vessels of the papillated structure can only be satisfactorily examined in properly injected preparations.

Some idea of the arrangement may be gained from the drawing given below.

Fig. 66.

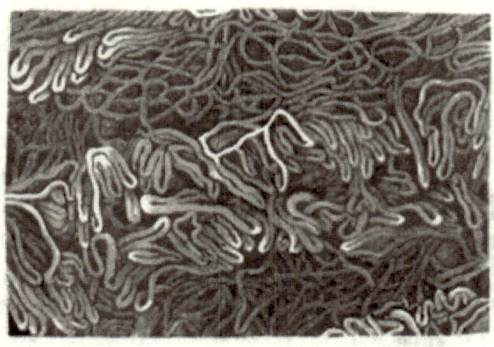

VESSELS OF THE PAPILLÆ OF ONE ENTIRE AND TWO HALF-RIDGES OF THE CUTIS.

The nerves of the skin may be examined in sections treated with dilute caustic soda or acetic acid, or a portion of skin may be boiled until it becomes transparent, and afterwards placed for a few hours in oil of turpentine, which renders the nerves white and shining. Fine sections must then be made with the double knife, and placed on the slide in a little oil of turpentine instead of water, covered with the piece of thin glass, and examined, first with a low, and afterwards with a high power.

The reticulated portion of the corium is closely connected to the papillary part, and contains the hair follicles and cutaneous glands, with collections of fat vesicles in its deepest portions. It is principally composed of white fibrous tissue, with a few elastic fibres, and in the neighbourhood of the hair follicles plain muscular fibres. All the elements may be studied by the ordinary process of teazing out portions of the corium in a little water on the slide, covering the specimen in the usual way, and examining it, first with the low, and afterwards with a higher power.

Smooth muscular fibre may be seen in successful sections of the skin of the scalp. The fibres arise from the superficial portion of the corium, and continue in the form of flattened bundles, obliquely towards the hair follicles.

The illustration given below will enable the student to recognise them when met with during the examination of various sections of the integument.

Fig. 67.

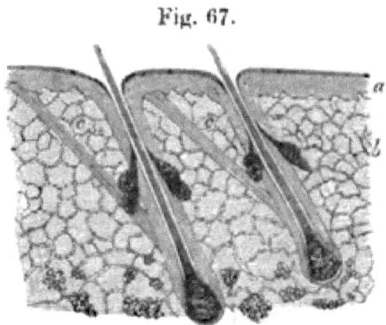

Perpendicular section through the scalp, with two hair-sacs. *a.* Epidermis. *b.* Cutis. *c.* Muscles of the hair-follicles.

Tela cellulosa subcutanea, or subcutaneous cellular tissue, is a compact membrane of connective tissue enclosing a large number of fat cells in its meshes in most parts of the body, and thus forming the '*Panniculus adiposus*,' or fatty layer, whose cells are arranged in variously sized masses or lobules, each lobule having a separate envelope of connective tissue.

No special directions are required for the examination of this structure, as it will be seen in most perpendicular sections of the skin.

Cutaneous Glands.

Irrespective of the lymphatics, which are tolerably abundant in some parts of the skin, that of the scrotum and round the nipple for example, there are certain glands and ducts specially belonging to the integument; these are the sudoriferous or sweat glands with their ducts, the sebaceous glands and ducts, and the ceruminous glands confined to the external auditory meatus.

Sudoriferous glands may be examined by making sections of slightly dried skin; that from the palm of the hand or sole of the foot is to be preferred.

The skin of the ball of the dog's foot or the udder of the mare will furnish large specimens.

After the section has been made with the razor, it must be soaked for a few hours in oil of turpentine, which will render it transparent. The preparation must next be transferred to the slide with a drop of turpentine, covered with the thin glass, and viewed with a low power.

The glandular coil will be seen, as indicated in the drawing, consisting of a single convoluted tube, in the interior of which the sweat is secreted.

Beautiful specimens may sometimes be obtained by using a solution of carmine in glycerine instead of turpentine. In that case, the section must be allowed to remain at least twenty-four hours in the coloured solution in order that the tissues may become sufficiently stained.

If the specimen be required for a permanent object, it must be put up in strong glycerine.

CUTANEOUS GLANDS.

Fig. 68.

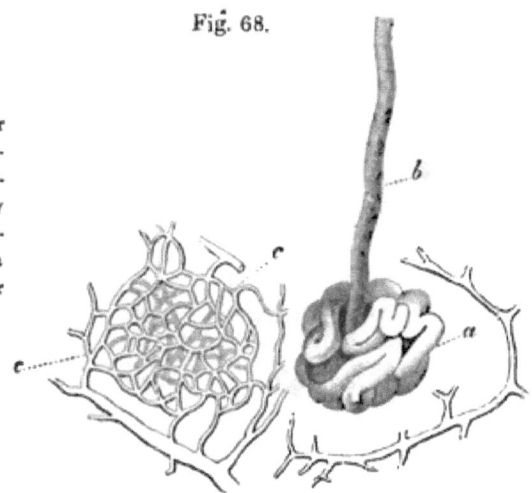

A sudoriferous glandular coil and its vessels, magnified 35 times. *a.* Glandular coil. *b.* Excretory duct, or sudoriferous canal. *c c.* Vessels of a glandular coil. After Todd and Bowman.

The vessels can only be seen in injected specimens.

The tube proceeding from the coil is nearly perfectly straight until it reaches the cuticle; through which it will be seen to take a spiral course as represented in illustration 69.

Fig. 69.

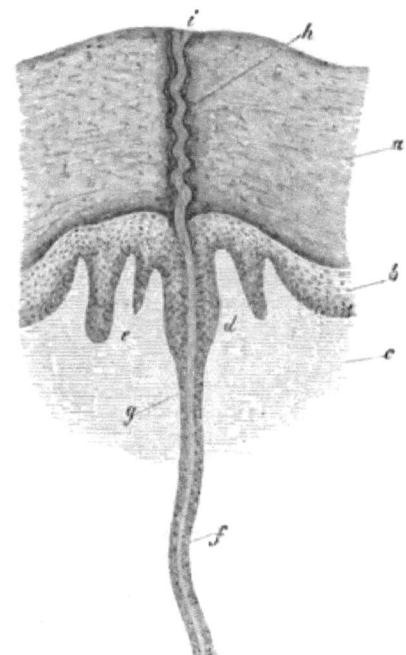

Perpendicular section of the epidermis and external part of the corium of the extremity of the thumb, carried transversely through two ridges, the preparation being treated with acetic acid and seen as magnified 50 times. *a.* Horny layer of the epidermis. *b.* Mucous layer. *c.* Corium. *d.* Simple papilla. *e.* Compound papilla. *f.* Epithelium of a sweat-duct passing into the mucous layer. *g.* Cavity of the same in the corium. *h.* In the horny layer. *i.* Sweat-pore.

Instead of oil of turpentine, a drop of acetic acid may be added to the section of skin, previously placed on the glass slide, the thin cover being immediately applied, and the specimen viewed as before with a high power. This method is more expeditious, and may be employed in conjunction with the former, on different sections, as the student must not expect to obtain a good view of the gland and its duct until he has examined several specimens.

The walls of the canals forming the sweat glands and ducts are formed of an external layer of fibrous tissue lined by a delicate basement membrane which supports the epithelium.

In the walls of the thickest tubes there is a middle layer of smooth muscular fibre.

In order to examine the minute structure of these canals, a single one must be isolated and viewed under a high power.

The drawing indicates what the observer should see.

Fig. 70.

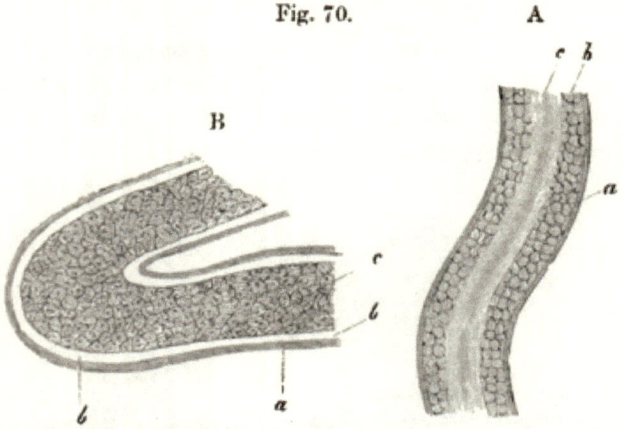

Sweat-ducts, magnified 350 times. A. One with thin walls and a central cavity without a muscular coat; from the hand. *a.* Connective investment. *b.* Epithelium. *c.* Cavity. B. A portion of a canal without a cavity, and with a delicate muscular layer; from the scrotum. *a.* Connective tissue. *b.* Muscular layer. *c.* Cells in the canal with yellow granules in their contents.

Sebaceous glands are small whitish bodies secreting an unctuous material and occurring principally in those parts of the skin which are covered with hair. They vary much in form, being either simple pear-shaped pouches, or simple racemose, or compound racemose glands; they generally open on the surface in conjunction with the hair-follicles.

SEBACEOUS GLANDS. 87

For the purpose of examining the sebaceous glands, sections of skin must be made as before directed, care being taken that they are not too fine. A drop of acetic acid or caustic soda will render the surrounding parts transparent, and permit the gland to be viewed under a low power.

The forms represented in the drawings may not all be discovered, but a careful examination of numerous specimens will usually lead to the observation of the principal varieties.

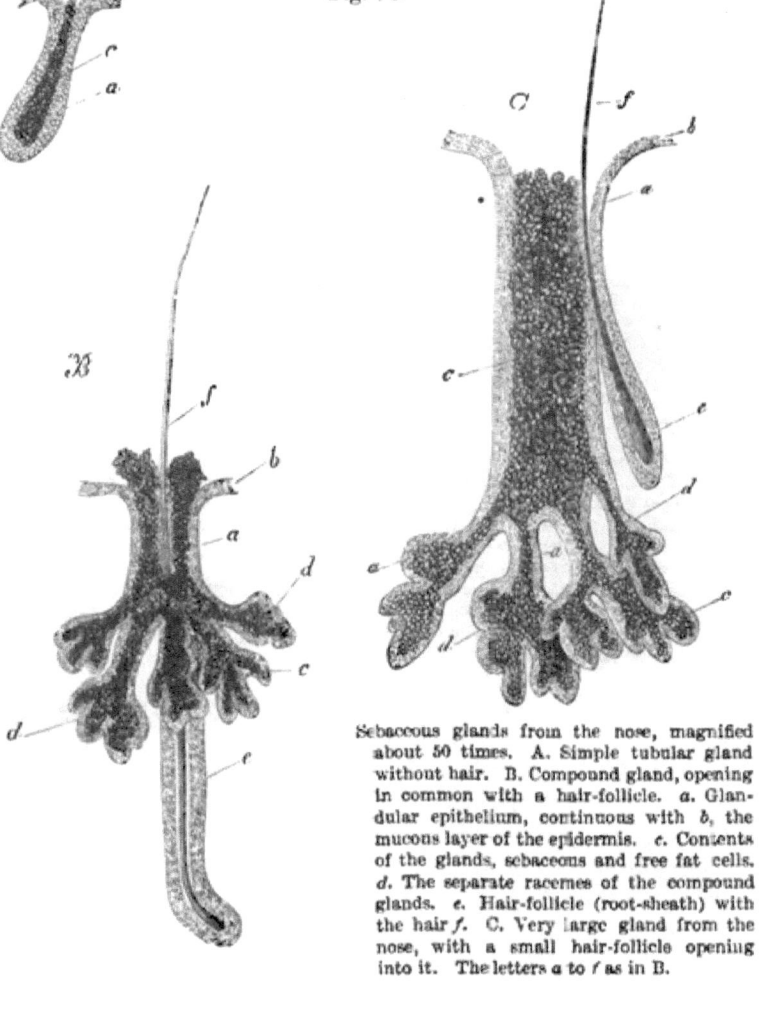

Fig. 71.

Sebaceous glands from the nose, magnified about 50 times. A. Simple tubular gland without hair. B. Compound gland, opening in common with a hair-follicle. *a.* Glandular epithelium, continuous with *b,* the mucous layer of the epidermis. *c.* Contents of the glands, sebaceous and free fat cells. *d.* The separate racemes of the compound glands. *e.* Hair-follicle (root-sheath) with the hair *f.* C. Very large gland from the nose, with a small hair-follicle opening into it. The letters *a* to *f* as in B.

The minute structure of the sebaceous glands may be studied in the skin of the scrotum, where they are easily isolated, or by macerating a portion of skin until the epidermis, with the hairs and cell masses of the glands, can be stripped off entire; by one of these methods, a single gland may be isolated and examined with a high power, after being placed on the slide in a little water; a little pressure on the covering glass will suffice to squeeze out some of the secretion, which may, however, be obtained in abundance from the surface of the skin of the sheep.

The drawing represents what the student should see, if he has succeeded in properly preparing a specimen of a gland.

Fig. 72.

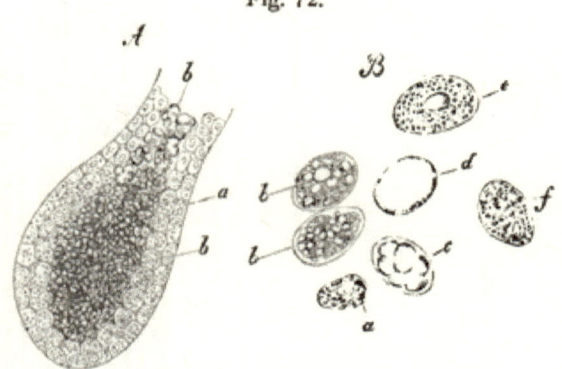

A. A glandular vesicle of an ordinary sebaceous gland, magnified 250 times. *a.* Epithelium sharply defined, but without being invested by a *membrana propria*, and passing continuously into the fat-cells in the interior of the gland-tube (the contents are rather indistinctly represented). B. Sebaceous cells from the gland-tubes, and the sebaceous matter, magnified 350 times. *a.* Smaller nucleated cells, containing but little fat, and possessing more the character of epithelium. *b.* Cells abounding in fat, without visible nuclei. *c.* Cells, in which the fat-particles are beginning to run together. *d.* Cell with one fat-drop. *e, f.* Cells whose fat has partly disappeared.

Each gland has an outer envelope of connective tissue proceeding from the hair-follicle, or from the corium in the case of glands not connected with hairs; within the envelope are several layers of nucleated epithelial cells, which gradually pass into cells containing numerous fat globules.

The ceruminous glands of the ear resemble the sudoriferous glands in structure, and are found between the integument and the cartilage of the external auditory meatus, forming a brownish layer. They may be examined by making sections through the skin of the meatus in the manner directed for the other parts of the integument.

The section should be placed on the slide in a little water, covered in the usual way, and examined with a low power; a drop of acetic acid will clear up the surrounding structures should they not be sufficiently transparent.

The drawing shows two of the glands as they should be seen.

Fig. 73.

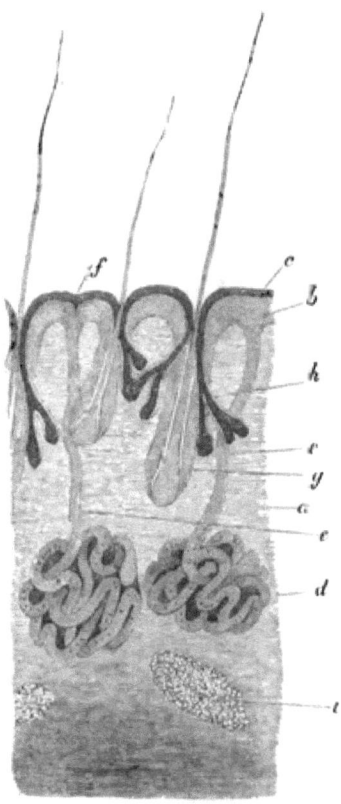

Section through the skin of the external auditory meatus, magnified 20 times. *a.* Corium. *b.* Stratum Malpighii. *c.* Horny layer of the epidermis. *d.* Coils of the ceruminous glands. *e.* Excretory duct of the same. *f.* Their openings. *g.* Hair-follicles. *h.* Sebaceous glands of the auditory meatus. *i.* Small collections of fat.

The cerumen is readily obtained from the meatus, and only requires to be distributed through a little water in the slide, covered with the thin glass, and viewed under a moderately high power. It will be found to contain a few hairs, epithelial cells, cells filled with fat, free fat globules, and some brownish granules.

HAIR.

Hair is composed, like nail, horn, and hoof, of modified epidermoid cells in various states of development. A hair consists of its free portion or shaft, and that part enclosed in the follicle—the root.

The root generally terminates in a bulbous swelling, in which is received the hair papilla at the bottom of the follicle as may be seen in a good specimen.

The illustration represents a perfectly successful preparation of the hair enclosed in its follicle.

Fig. 74.

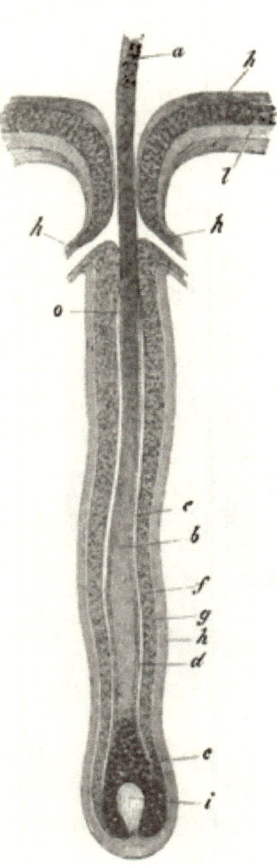

Hair and hair-follicle of middling size; magnified 50 times. *a.* Shaft. *b.* Root. *c.* Bulb. *d.* Epidermis of the hair. *e.* Inner root-sheath. *f.* Outer root-sheath. *g.* Structureless membrane of the hair-follicle. *h.* Transverse and longitudinal fibrous layer of the same. *i.* Papilla of the hair-follicle. *k.* Excretory ducts of two sebaceous glands. *l.* Cutis. *m.* Mucous, and *n*, horny layer of the epidermis, the latter entering a certain way into the follicle. *o.* End of the inner root-sheath.

Microscopic Examination of Hair.—In the first place, a single hair from the head may be taken, placed in a little water on the slide, covered in the usual way, and examined first with a low, and afterwards with a high power.

The external surface will be found to be covered with plates or scales, as represented in the drawing.

Fig. 75.

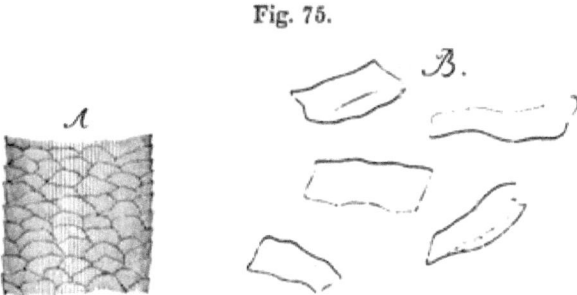

A. Surface of the shaft of a white hair, magnified 160 times. The curved lines designate the free border of the epidermic plates. B. Epidermic plates from the surface, isolated by caustic soda, and magnified 350 times. Either one or both of their borders are folded, and consequently appear dark.

In consequence of the reflection of the light from the centre of the hair, as from any other cylindrical body, it appears to be hollow or tubular, but the examination of transverse sections will prove this appearance to be fallacious, excepting in some kinds of hair to be afterwards described.

Transverse sections may be made in a variety of ways. The most simple plan is to take a bundle of hairs and dip them in hot liquid gelatine, and allow them, after being removed from the liquid, to dry slowly in the air until sufficiently firm to be cut with a razor. Very fine sections may then be made and thrown into warm water, which again dissolves the gelatine and allows the sections to fall to the bottom; some of these are to be collected and placed on the slide with a little water, covered and examined in the usual way.

Good specimens should show the parts represented in the illustration (fig. 76).

Fig. 76.

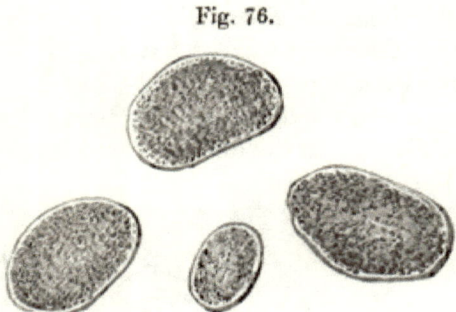

Transverse section of human hair.—High power.

The dark lines surrounding the sections represent the cut edge of the layer of imbricated scales. The dots are the ends of the fibres which are seen to be less numerous near the centre of the section than towards the circumference.

Sections of the beard are readily made by shaving twice at short intervals, washing the lather from the razor, and afterwards collecting the sections as before.

Longitudinal sections may be made by scraping or splitting a hair on the glass slide with a sharp scalpel or razor.

In examining entire hairs or sections, a drop of liquor potassæ allowed to flow under the edge of the covering glass will render the structures more distinct, more particularly the scales of the cortex.

To demonstrate the fibrous structure which constitutes the principal bulk of the hair, it is necessary to boil it in sulphuric acid for ten minutes, or to macerate for a few hours in caustic potass. The hair so treated splits up into minute fibres, some of which should be placed in a drop of water on the slide, covered in the usual way, and examined with a high power.

If the fibres are not sufficiently separated, they may be teazed out and again examined; when properly prepared they will be seen to consist of fusiform cells containing colouring matter, and sometimes nuclei.

The illustration (fig. 77) shows a preparation obtained by the action of acetic acid instead of sulphuric.

STRUCTURE OF HAIR.

Fig. 77.

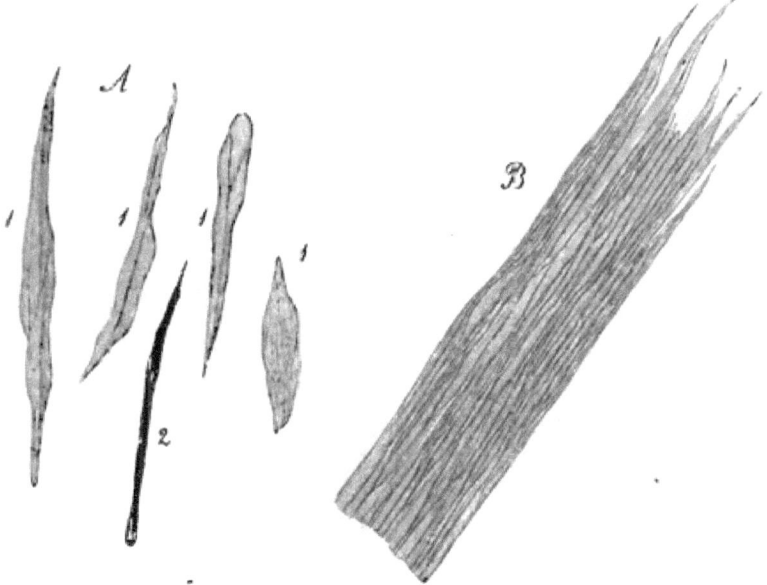

Plates or fibre-cells of the cortical substance of a hair, treated with acetic acid; magnified 350 times. A. Isolated plates: 1. From the surface (three single and two united). 2. From the side. B. A bundle composed of many such fibre-cells.

The medulla or pith is the central line of cells extending from the bulb to near the apex, but generally absent in fine hairs and in coloured hairs from the head. It may be examined by steeping white hairs in caustic potass until they swell up, and then compressing one of them on the glass slide, and viewing it with a high power; the cells will be seen to be transparent, rectangular in form, and occasionally furnished with dark fatty-looking granules.

The follicle which receives the root of the hair sometimes reaches down to the subcutaneous fatty tissue when the hair is of some length and thickness. It consists of an external fibrous coat contiguous with the corium, and an epidermic lining extending from the cuticle.

The external coat is composed of two distinct fibrous layers, the inner one only reaching as high as the entrance of the sebaceous glands, and a delicate basement membrane.

94   HISTOLOGY.

Hairs taken from various animals form interesting objects of study.

The methods of preparation do not differ in any way from those suggested for the examination of human hair.

Some of the peculiarities observable upon the examination of different hairs are illustrated in the annexed drawings.

Fig. 78.

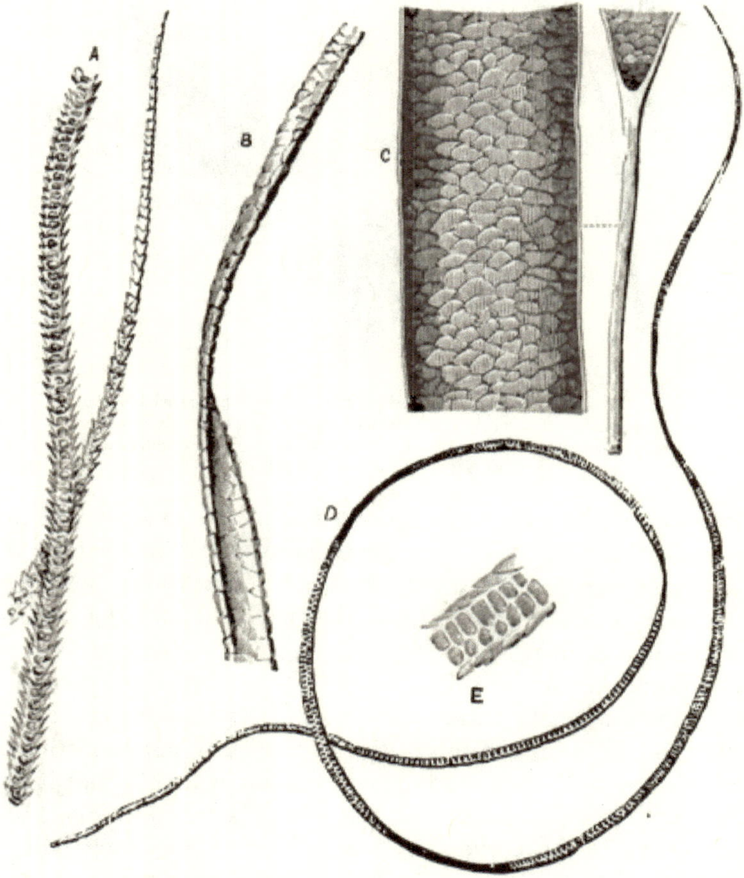

A. Jointed hairs from the Indian bat.
B. Hairs of the flying fox, showing imbricated scales.
C. Hair from the pronged buck of California, having its surface covered with tesselated epithelium.
D. Hair from the white rat, showing numerous air cavities in the centre. A portion in middle, E, is more highly magnified.

HAIRS OF VARIOUS ANIMALS. 95

The transition from the solid to the tubular form of hair is beautifully seen in the hairs of certain animals, as illustrated in the accompanying drawing of sections.

Fig. 79.

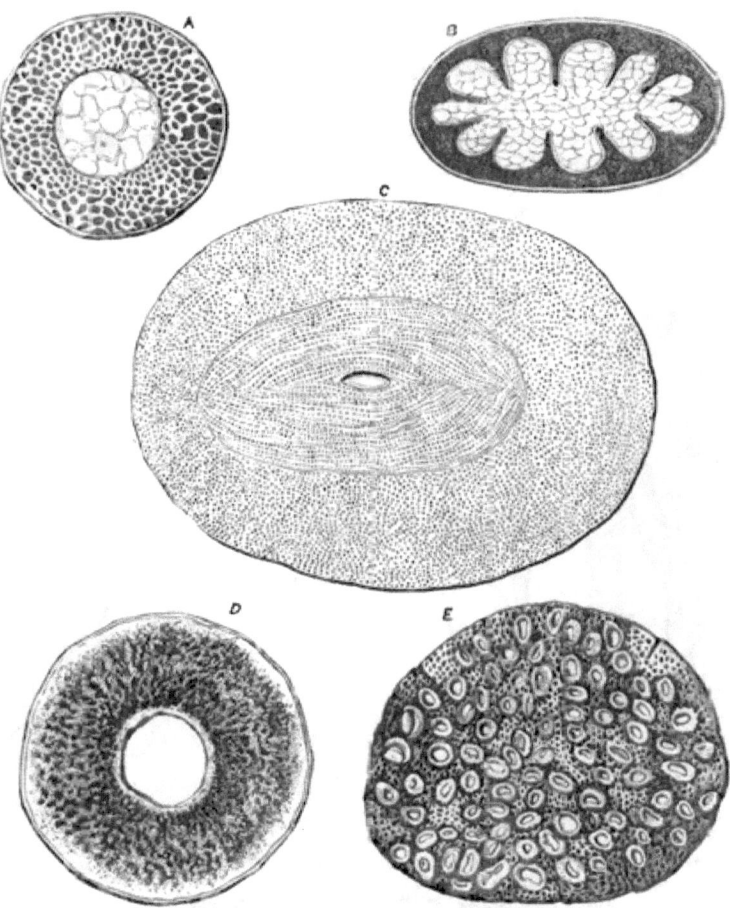

A. Hairs from the ant-eater, showing the central part filled with pith.
B. Hair of peccary, with greater development of the pith.
C. Section of hair of walrus, indicating the commencement of the tubular form.—Magnified 100 diameters.
D. Section of the whisker of the tiger, in which the tube is perfect.
E. Section of hair of an elephant, showing the combination of a number of tubes united together, exactly resembling the arrangement of the hoof-horn, another cutaneous appendage.—Magnified 50 diameters.

96 HISTOLOGY.

Hairs and hair-like bodies will often occur as accidental objects upon the slide, and may give rise to either serious or ludicrous mistakes, in proof of which it is only necessary to allude to a recent instance of a hair from a rat having been sent for examination, as a new form of entozoon found in the secretion from the ear.

For obvious reasons the objects most commonly present accidentally, are fibres of wool from flannel, and fibres of cotton, linen, or silk, with the characters of which the student should be familiar, in order to avoid confounding them with any other structures under examination. Illustrations of these objects are given, as a guide to assist the observer in detecting them when present.

Fig. 80.

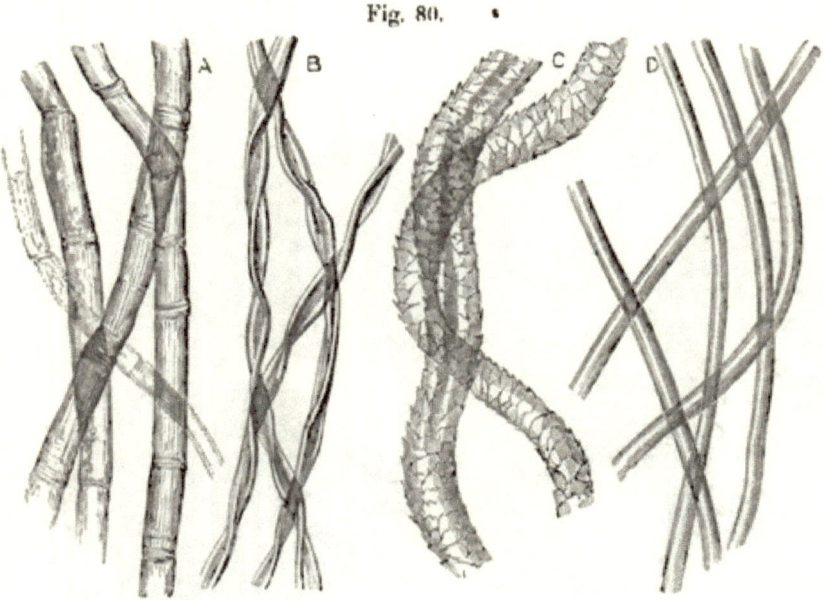

A. Linen fibres, solid, round, and tapering, not twisted, and possessing no medulla.
B. Cotton fibres, flattened and twisted tubes, having an indistinct medulla.
C. Sheep's wool, distinguished by the covering of imbricated scales and generally the absence of medulla.
D. Silk fibres, round and solid, without any medulla; occasionally a few twisted fibres are seen, but they are easily distinguished from the flattened tubes of cotton.

OF THE NAILS.

Nail is nothing more than epidermis, the elements of which have assumed the character of horn.

A nail is divided into three parts: 1. root; 2. body; 3. edge.

The root is covered by a fold of skin, immediately in front of which is a white portion, resembling a half-moon in shape, and termed the lunula.

Matrix is that part of the cutis vera which secretes the nail. It differs but little from the papillated structure of the corium in other parts, except that under the body of the nail it is very dense, and is arranged to form ridges or lamellæ, which correspond to the under surface of the nail.

Like the epidermis, nail is composed of two layers, an upper hard or horny surface, and a lower softer stratum (stratum Malpighii).

The position of these parts can only be seen in good sections, to make which some care is necessary. First a nail with its bed or matrix must be separated from the bone of the finger, and dried until sufficiently firm to be sliced with the razor.

Transverse and longitudinal sections may be made, and on being placed in a little water on the slide, they soon swell up and resume their natural characters.

Under a low power the appearances shown in the illustration will be recognised, but the minute structure of nail cannot be perfectly seen without the aid of chemical agents.

Fig. 81.

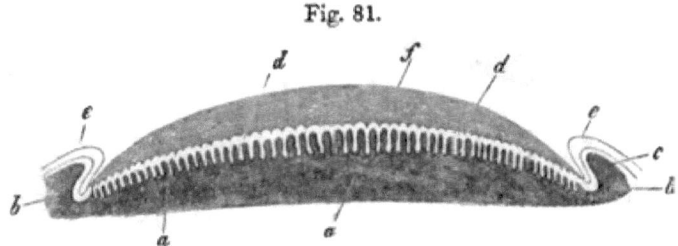

Transverse section through the body and bed of the nail; magnified 8 times. *a.* Bed of the nail, with its ridges (black). *b.* Corium of the lateral parts of the wall of the nail. *c.* Stratum Malpighii of the same part. *d.* Stratum Malpighii of the nail, with its ridges (white). *e.* Horny layer on the wall of the nail. *f.* Horny layer of the nail, or proper nail substance, with shallow notches upon its under surface.

If a portion of the section should happen to be sufficiently thin, a high power will show the two layers of which the nail is composed, as in the next drawing.

Fig. 82.

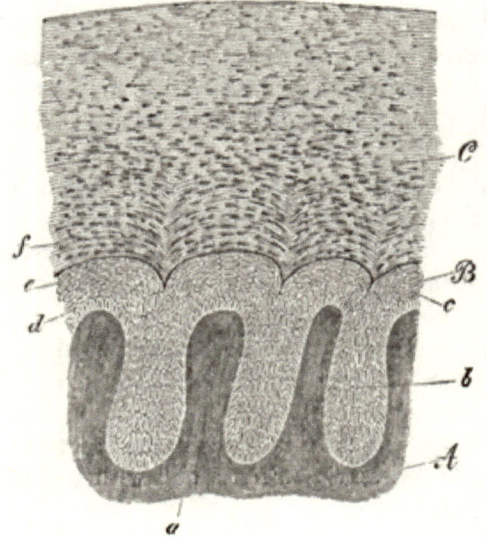

Transverse section through the body of the nail; magnified 250 diameters. A. Cutis of the bed of the nail. B. Mucous layer of the nail. C. Horny layer of it, or proper nail-substance. *a.* Laminæ of the bed of the nail. *b.* Laminæ of the stratum Malpighii of the nail. *c.* Ridges of the nail itself. *d.* Deepest elongated cells of the mucous layer of the nail. *e.* Upper flat-cells of it. *f.* Nuclei of the proper substance of the nail.

Longitudinal sections are more difficult to make and are not so instructive as transverse ones. The parts seen in the next figure will rarely be obtained perfectly in any one specimen.

Fig. 83.

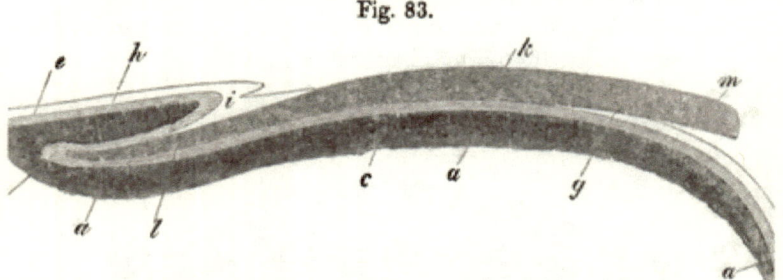

Longitudinal section through the middle of the nail and its bed; magnified 8 times. *a.* Bed of the nail, and cutis of the back and point of the finger. *b.* Mucous layer of the point of the finger. *c.* Of the nail. *d.* Of the bottom of the fold of the nail. *e.* Of the back of the finger. *f.* Horny layer of the point of the finger. *g.* Beginning of it under the edge of the nail. *h.* Horny layer of the back of the finger. *i.* Termination of it upon the root of the nail. *k.* Body. *l.* Root. *m.* Free edge of the proper substance of the nail.

STRUCTURE OF NAIL.

The minute structure of nail, of which very little is usually seen in section, may be studied in specimens that have been soaked for some hours in caustic potash. A scraping of the softened nail should be placed in a drop of water on the slide, teazed out, and covered in the usual way.

Under a high power very delicate nucleated cells will become apparent, as represented in the woodcut.

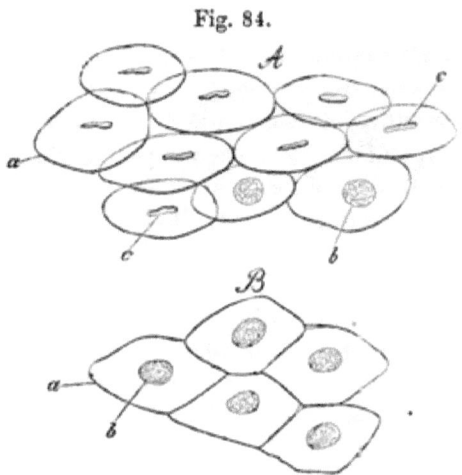

Fig. 84.

Nail-plates, boiled with caustic soda; magnified 350 times. A. Viewed from the side. B. From the surface. *a*. Membranes of the distended elements. *b*. Their nuclei from the surface. *c*. from the side.

The cells forming the nail may also be rendered apparent by allowing a drop of caustic potash to flow under the edge of the covering glass on to a fine section under examination. By this method the arrangement of the cells in layers will be clearly seen.

During the growth of the nail it appears that the soft or mucous layer (with its ridges) remains in its position, while the upper or horny layer is continually pushed forward from the root.

The growth of a nail from root to edge has been found to occupy twelve weeks.

HORN AND HOOF.

Both these structures are composed of the same elementary tissues as nail and epidermis.

The method of examination is in every respect the same as that recommended for the study of nail. Portions of hoof and horn, with the soft structure attached, are to be dried, and afterwards cut with a razor, and the sections treated in the usual manner. To render the cells distinct, caustic potass is necessary, and must be employed in the way previously directed.

Horn is as nearly as possible identical in structure with nail, as will be seen by making a fine transverse section of a portion of the horn of an ox, and comparing it with the drawings of transverse sections of the human nail (figs. 81, 82, pp. 97, 98).

Some of the cells of the deep or mucous layer of horn contain pigment, which is also frequently found in the same part of human nail; in that of the negro it is invariably present.

The ridges at the under surface of horn are not so close together nor so well developed as those of the nail; but with these unimportant differences, the resemblance is perfect.

Hoof.—The structure of hoof, as seen in sections, is tubular, the canals running from above downwards.

In transverse sections of hoof, the openings of the numerous canals will be observed surrounded by concentric lines, indicating the layers of compressed epithelial cells. A drop of caustic potass placed at the edge of the covering glass, will, in a few minutes, cause the cells to expand without destroying their concentric arrangement.

In the embryo hoof the cells are seen very distinctly surrounding the canals, looking remarkably like the adult hoof after the addition of a drop of caustic potass.

The next drawing represents transverse sections of the hoof in its natural state, and after treatment with potass.

Fig. 85.

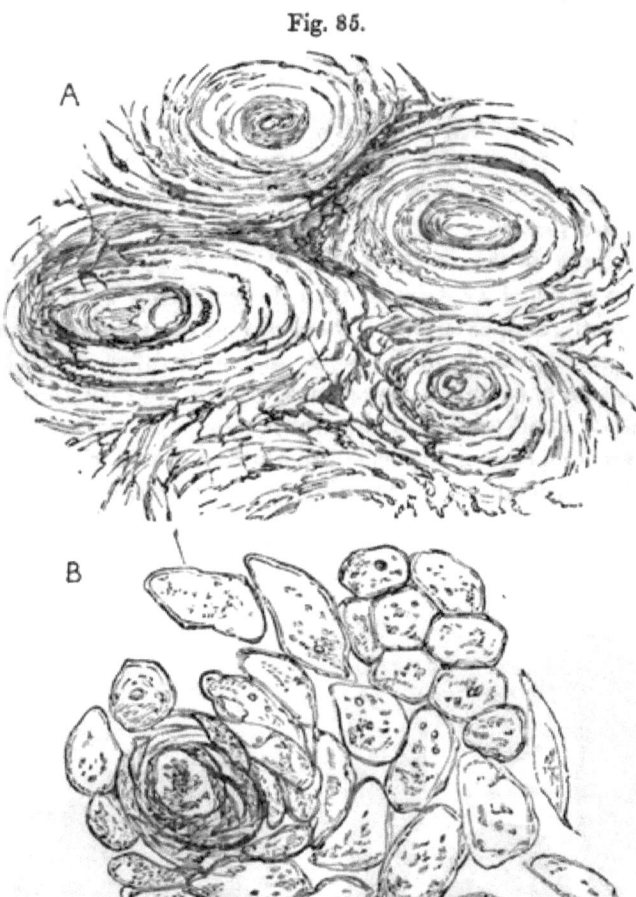

Transverse section of horse's hoof. A. Natural condition. B. After the addition of potash.—Magnified 200 times.

Specimens of nail, horn, or hoof, may be mounted dry, or soaked in turpentine, and afterwards put up in thin Canada balsam.

TEETH.

The several structures of which a tooth is composed are arranged in definite order round a central space, which is termed the pulp-cavity, containing a soft reddish substance which abounds in nerves and vessels.

The bulk of the tooth is formed of ivory, or dentine, over which is reflected a coating of enamel, principally upon the crown of the tooth, and, external to these structures, the cementum, or crusta petrosa, which, in the horse and some other of the herbivora, forms a complete investiture, but in man only covers the fang and a small portion of the crown.

The accompanying sections of the human molar will convey an idea of the arrangement of the different structures, and the position of the pulp cavity.

Fig. 86.

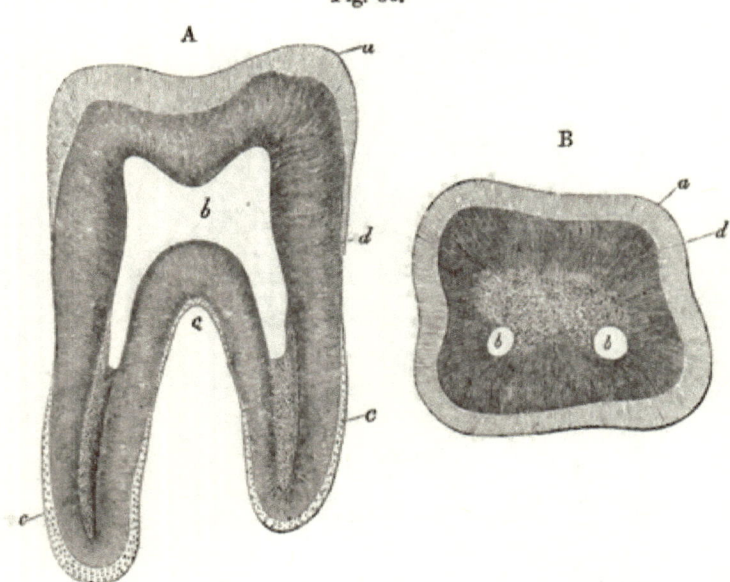

Human molar tooth; magnified about 5 times. A. A longitudinal, B. a transverse section. *a.* Enamel. *b.* Pulp-cavity. *c.* Cement. *d.* Dentine, with the canaliculi.

The best view of the three tissues, ivory, enamel, and crusta petrosa, may be obtained by examining a section of a tooth of a horse.

The drawing representing these parts is taken from a very successful preparation of an incisor tooth of that animal.

Fig. 87.

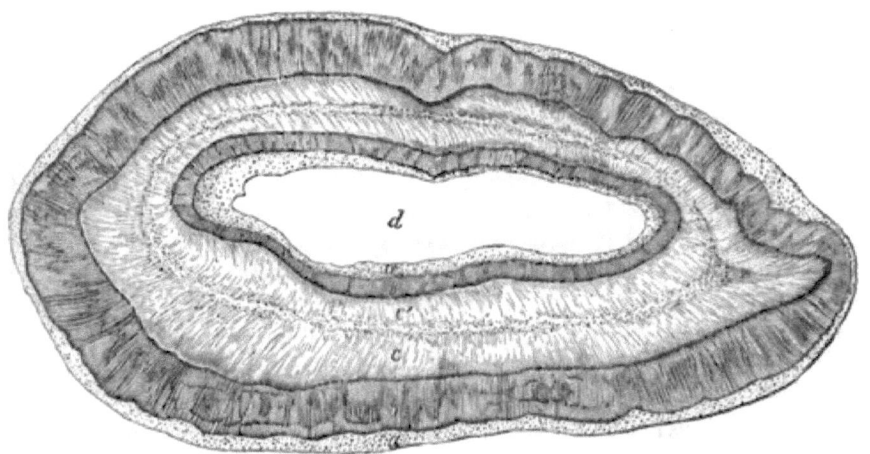

Section of incisor of horse. *a.* Crusta petrosa. *b.* Enamel. *c.* Dentine. *d.* Central cavity or infundibulum, formed by an inversion of the structures which consequently occur in a reverse order.—Low power.

Microscopic examination of Tooth.—In the first place sections of softened teeth may be studied, these are prepared by macerating a tooth in dilute acid as directed for bone, and afterwards cutting fine sections with a razor.

Sections of fresh teeth will also be necessary, but their preparation is so tedious that they are better purchased ready for examination, at prices varying from 1*s.* to 5*s.*, according to their quality.

A variety of sections will be necessary, in order to demonstrate the characters belonging to the different structures.

Very valuable and instructive preparations may be made by Dr. Beigel's method of using acids.

If a tooth be boiled for a few minutes in sulphuric acid, the cementum, cuticle of the enamel and the ivory are all dissolved, leaving the enamel only intact in the form of a small horse-shoe.

By arresting the process at the proper moment it is possible to obtain specimens of enamel and ivory together, the cementum only being dissolved out by the acid.

Nitric acid, on the other hand, destroys the enamel before it attacks the other structures.

Dentine.—In a longitudinal section of a tooth the dentine appears to be transparent; under a microscope it is seen to consist of tubes running from the centre to the external surface, and describing two or three spiral curves in their course (primary curves), as seen in fig. 86, A. The tubes are also finely undulated throughout their entire length, having as many as 200 flexures in a line (secondary curves).

The dentinal canals are largest at their commencement from the pulp cavity, becoming finer as they proceed outwards, and terminating, in some parts of the tooth, in small irregular cavities like lacunæ.

From various parts of the dentinal tubes small branches arise, which ramify and anastomose in all directions.

In the drawing all these peculiarities of the structure of dentine are represented.

Fig. 88.

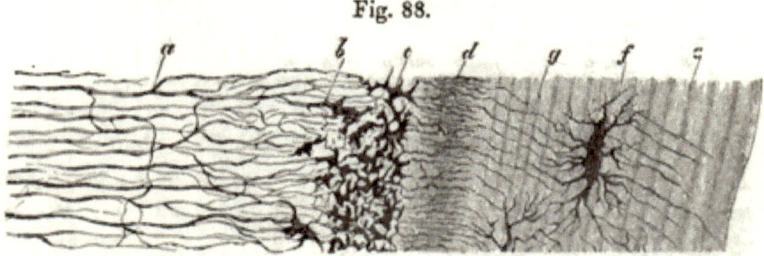

Dentine and cement, from the middle of the fang of an incisor tooth. *a.* Dentinal tube. *b.* Interglobular spaces, looking like lacunæ. *c.* Finer interglobular spaces. *d.* Commencement of the cement, with numerous closely-disposed canals. *e.* Lamellæ of the same. *f.* Lacunæ. *g.* Canaliculi.—Magnified 350 times. Of man.

Other sections are necessary to enable the student to see the tubular character of dentine, as in the subjoined woodcut, which represents a transverse section of the human molar, showing the homogeneous matrix containing the dentinal tubules.

Fig. 89.

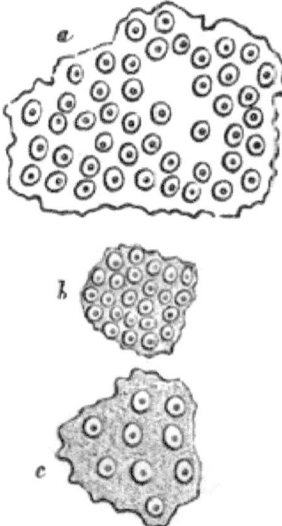

Transverse section of dentinal canals, as they are usually seen. *a.* Ordinary distance apart. *b.* More crowded. *c.* Another view. Human molar.—Magnified 400 times.

The structure of dentine in the tooth of the saw-fish is distinguished by the presence of numerous large lacunæ, with radiating and tortuous canaliculi, as in reptiles' bone.

Fig. 90.

Section of tooth of saw-fish.—Magnified 200 times.

Osteo Dentine is the name given to a substance found in old teeth, filling up the pulp cavity, either wholly or partially. This structure partakes of the characters of dentine and bone, having canals surrounded by concentric lamellæ like Haversian canals; from these canals numerous tubules ramify in all directions, in the same manner as the tubules of dentine.

Enamel is the hardest and most brittle of the tooth structures. It is composed of prisms about $\frac{1}{3500}$ of an inch in diameter, arranged closely together in slightly waving lines, sometimes crossed by transverse striæ, as indicated in the next drawing, representing a longitudinal section of enamel.

Fig. 91.

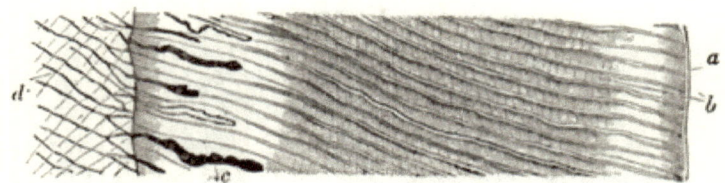

Human dentine and enamel; magnified 350 times. *a.* Enamel-cuticle. *b.* Enamel-fibres, with fissures between them and transverse lines. *c.* Large cavities in the enamel. *d.* Dentine.

The ends of the enamel prisms can be seen in good transverse sections; in some parts of the preparation a side view will generally be obtained, in consequence of the oblique course of the fibres.

In the illustration the prisms are represented quite close together, but occasionally minute spaces or canals are found to exist between them.

Fig. 92.

Surface of the enamel, with the extremities of the enamel-fibres; magnified 350 times. Of the calf.

Cementum, or crusta petrosa, is the only one of the tooth structures which resembles bone, possessing, as it does, lacunæ and canaliculi, and occasionally Haversian canals.

Sections are easily obtained, but it is only rarely that any one specimen presents all the characters of bone.

The illustration represents the structure as seen in a favourable specimen.

Fig. 93.

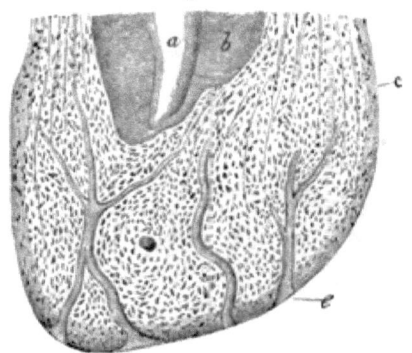

Cement and dentine of the fang of an old human tooth. *a.* Dental Cavity. *b.* Dentine. *c.* Cement with lacunæ. *e.* Haversian canals.

Dental Pulp, contained in the cavity of the tooth, may be obtained for examination by breaking a tooth in a vice. It is a soft reddish substance, intimately connected to the inner surface of the dentine, and composed of connective tissue, with round or elongated nuclei, and numerous nerves and vessels, which can only be satisfactorily studied in finely injected specimens.

There is also a fluid like mucus, coagulable by acetic acid; and, lining the pulp cavity, or resting on the surface of the pulp, are numerous cylindrical or conical nucleated cells, arranged perpendicularly to the surface, like columnar epithelium.

The vessels of the dental pulp are represented in the drawing, taken from a very beautiful injection of a molar tooth of a cat. See Frontispiece.

DIGESTIVE CANAL.

The mucous membrane and the numerous glandular structures of the digestive system furnish interesting and important objects for microscopic observation.

Sections of the membrane are to be made from dried preparations, or from specimens that have been hardened in alcohol.

For the purpose of examining the glands of the oral cavity, the sections should be continued through the submucous tissue.

The illustrations show two kinds of glands, viz.: racemose

Diagram of two ducts of a lobule of a mucous gland. *a.* Excretory duct of the lobule. *b.* Lateral branch. *c.* The *acini* of that branch *in situ*. *d.* The same spread out, and the duct displayed.

glands from the mouth, and a follicular gland from the root of the tongue; these last are identical in structure with the follicles united to form the tonsils.

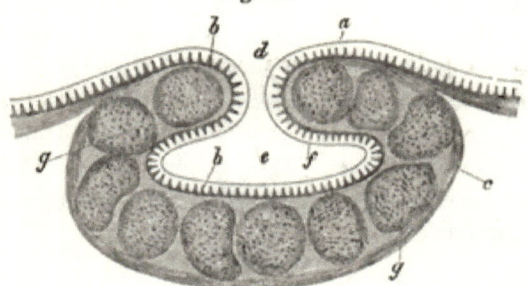

Follicular gland, from the root of the tongue of man. *a.* Epithelium lining the same. *b.* Papillæ. *c.* Outer surface of the gland, with the envelope of areolar tissue. *d, e.* Cavity of the gland. *f.* Epithelium of the same. *g.* Follicles in the thick wall of the gland.—Magnified 30 times.

Each follicular gland is a thick-walled capsule lined by a continuation of the membrane of the mouth.

Between the mucous lining and the external coat there are several small capsules, somewhat resembling the solitary glands.

All these structures are best examined in injected preparations.

The surface of the tongue is covered with numerous papillæ, of which there are three distinct varieties, as indicated in the drawing.

Fig. 96.

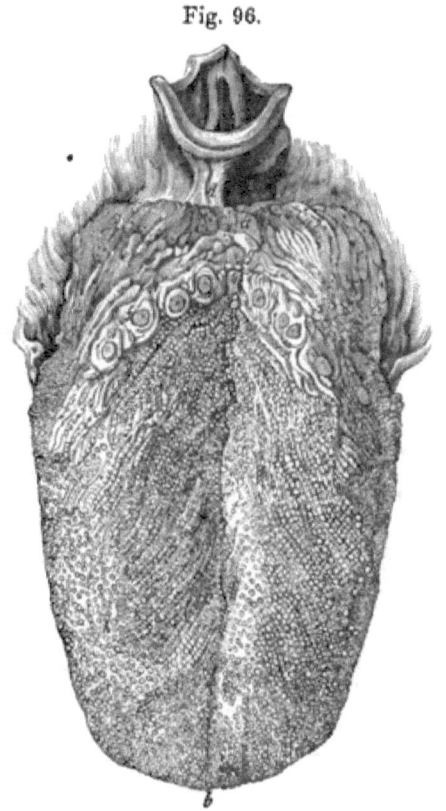

Upper surface of human tongue. *f.* Papillæ circumvallatæ. *e.* Non-papillated portion of tongue, containing the follicular glands.

In front of the circumvallate papillæ are seen the numerous filiform papillæ, with a few fungiform scattered about the posterior part and on the edges and tip of the tongue, distinguished by their spheroidal shape. All the rest are coniform and filiform.

From the fresh tongue single papillæ may be snipped off and examined in the usual manner.

Sections may be made with a razor from a tongue that has been partly dried, or boiled hard.

These are to be placed on a slide in a drop of water, and covered with a piece of thin glass. A drop of caustic soda will assist to render the nerves and epithelial cells apparent.

Bloodvessels can only be well seen in transparent injections.

Illustrations are given of the three classes of papillæ, for the purpose of enabling the student to test the correctness of his observations.

Fig. 97.

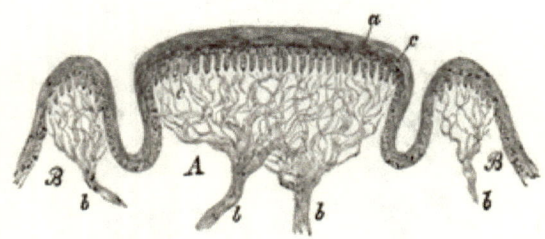

Papilla circumvallata of man in section. A. Proper papilla. B. Wall. *a*. Epithelium. *c*. Secondary papillæ. *b, b*. Nerves of the papillæ and the wall.—Magnified about 10 times.

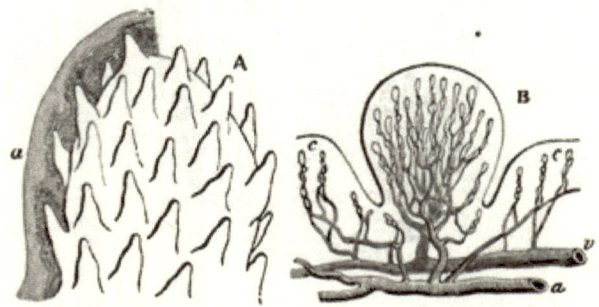

A. Fungiform papilla, showing the secondary papillæ on its surface, and at *a* the epithelium covering them over. Magnified 35 diameters. B. Another, with the capillary loops of its secondary papillæ injected. *a*. Artery. *v*. Vein. The groove around the base of some of the fungiform papillæ is here represented, as well as the capillary loops, *c c*, of some neighbouring simple papillæ.—Magnified 18 diameters. After Todd and Bowman.

Fig. 97—*continued*.

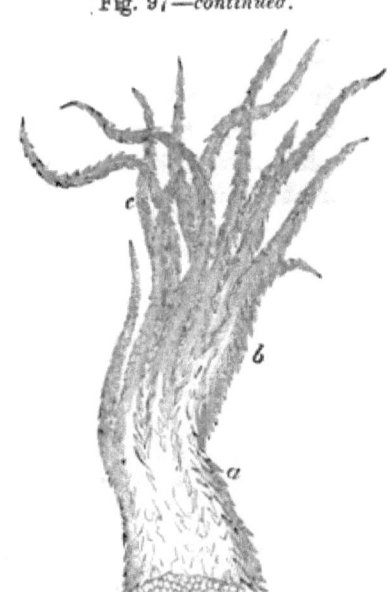

A filiform papilla taken from the dorsum of a tongue in which the fur was much developed; magnified 30 diameters. *a.* Imbricated scaly epithelium investing the cylindrical portion of the papilla. *b.* The commencement of its breaking up. *c.* Its separation into its ultimate filamentary processes. *d.* The deep layer of epithelium exposed by the removal of the more scaly superficial one.

The filiform papillæ are largest in the neighbourhood of the circumvallatæ, but they are less in diameter, though longer, than the other two kinds, and present greater diversity of form.

The Salivary Glands.—For the purpose of investigation the student may select a portion of the parotid, which he will find to consist of minute lobules, united to form larger divisions, which are again united to constitute the gland.

By teazing out a small fragment in a drop of water on the slide, the various elements will be seen, viz.: the ducts, with the epithelial lining, fat granules, and connective tissue.

Saliva requires no preparation; a drop being placed on the glass slide should be covered with the thin glass cover and

112 HISTOLOGY.

viewed with the high power. It will be seen to consist of a clear fluid, containing various accidental constituents, such as epithelial cells and mucous corpuscles.

The mucous membrane of the œsophagus, particularly at its lower third, furnishes numerous examples of mucous glands, which are beautifully seen in injected specimens, one of which has been selected for the next illustration.

Fig. 98.

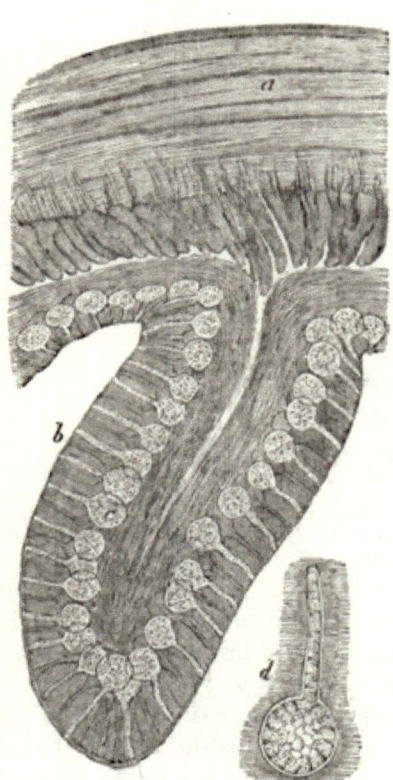

Œsophageal glands (human). *a.* Wall of œsophagus. *b.* Fold of mucous membrane. *c.* Gland follicle. *d.* Same more highly magnified.

The Mucous Membrane of the Stomach.—For the purpose of examination, the fresh stomach of a pig should be obtained and cut open.

Preparations of the tubular glands or follicles, which secrete the mucus that protects the stomach from the action of the gastric juice during digestion, may then be made by stretching a portion of the viscus taken from the pyloric end over a piece of cork, and cutting fine sections under water, by means of Valentin's knife.

The specimen should be floated on to the slide and covered in the usual way. Viewed under a low power the follicles should appear as in the illustration, but if the section is too thick, it will be necessary to teaze it out slightly before the glands can be seen. A portion of the stomach may be left on the cork for a day or two to become dry, after which sections may be cut with a razor, and placed in a little water on the slide, when they will soon recover their natural condition, and may be examined with a low power.

Fig. 99.

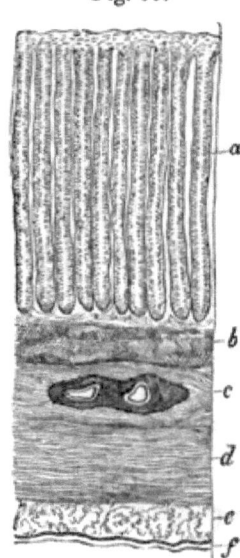

Perpendicular section through the coats of the stomach of the pig, from the pylorus; magnified 30 times. *a.* Glands. *b.* Muscular layer of the mucous membrane. *c.* Submucous tissue (*tunica nervea*), with cut vessels. *d.* Transverse muscular layer. *e.* Longitudinal muscular laminæ. *f.* Serous membrane.

The peptic glands supposed to secrete the gastric juice, are found in the cardiac portion of the stomach; the distinction between these and the tubes of the pyloric portion is very marked, as will be seen by reference to the illustration, in which the two forms are placed side by side.

Fig. 100.

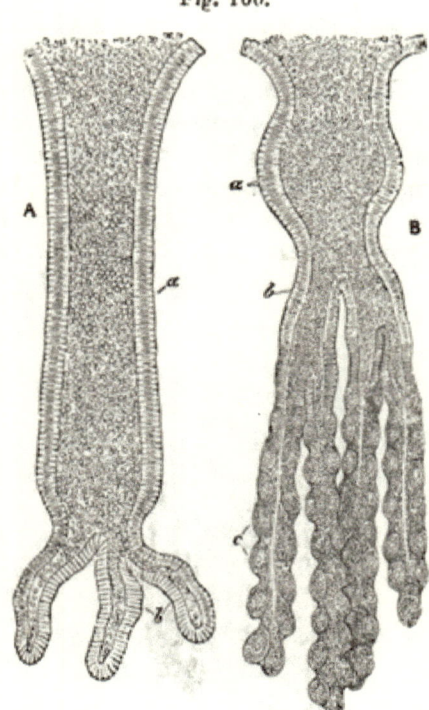

Tubes from the cardiac and pyloric regions of the dog's stomach, to show the contrast of their structures; magnified 60 diameters. Altered from Kölliker. A. Pyloric tube. *a.* Primary tube. *b.* Three secondary tubes. B. Cardiac tube. *a.* Primary tube lined by columnar epithelium. *b.* Two secondary tubes. *c.* Four terminal branches containing large oval cells.

Mucous Membrane of the Small Intestines.—Sections of the intestine are to be made in the manner directed for the stomach, and prepared on the slide in the usual way. The arrangement of the various coats, and also the villi and Lieberkühn's follicles, can be seen under a low power, as shown in the drawing.

GLANDS OF THE INTESTINES.

Fig. 101.

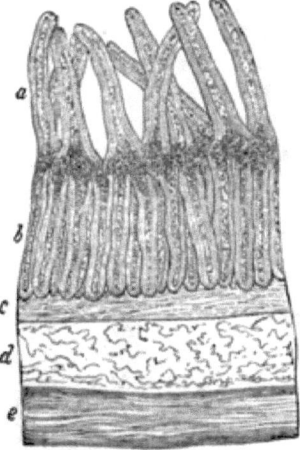

Vertical and longitudinal section of the small intestine in the lower part of the jejunum, showing the general arrangement of its coats; magnified 50 diameters. *a.* Villi. *b.* Intestinal tubes. *c.* Submucous areolar tissue. *d.* Circular fibres of the muscular coat. *e.* Longitudinal fibres, external to these, covered by peritoneum.

The villi can be seen also, with the unaided eye, on a portion of intestine placed under water.

The next drawing represents the vessels of two villi injected.

Fig. 102.

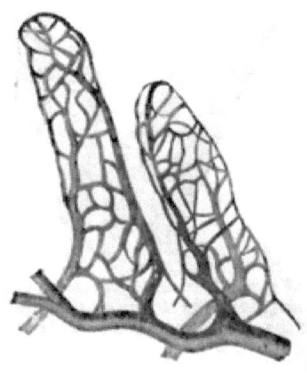

Vessels of two of the *villi* of the mouse, from an injection by Gerlach. Magnified 45 times.

116   HISTOLOGY.

For examination under a high power, one or two villi may easily be snipped off a portion of intestine, and placed on the glass slide in the usual way.

Enclosed in the villi, which are simply prolongations of mucous membrane, there are bloodvessels, lacteals, and muscular cells.

The lacteals are best examined in sections made with fine scissors, and treated, when under the microscope, with caustic soda. The preparation, if successful, will present the apparancees seen in the illustration.

Fig 103.

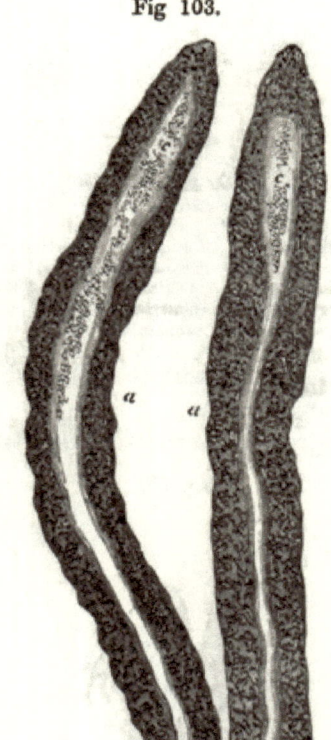

Two villi, denuded of epithelium, with the lacteal vessel in their interior. From the calf; magnified 350 diameters. After Kölliker. *a.* Limitary membrane of the villus. *b.* Matrix or basis of the same. *c.* Dilated blind extremity of the central lacteal. *d.* Trunk of the same.

Brunner's glands, found in the sub-mucous tissue of the duodenum, may be exposed by dissecting the peritoneal and muscular

GLANDS OF THE INTESTINES. 117

coats from a portion of intestine near the pylorus, previously distended with air. They are readily distinguished as yellow rounded or flattened bodies, lying beneath the membrane.

Sections of the fresh intestine, or of a portion that has been boiled in acetic acid, and even sections of a dried preparation, may be examined to see the relation of these glands to the surrounding tissue.

If the dried intestine is employed, a drop of caustic soda is to be added when the specimen is under the microscope.

The drawing represents one of the glands, as seen in a section under a low power.

Fig. 104.

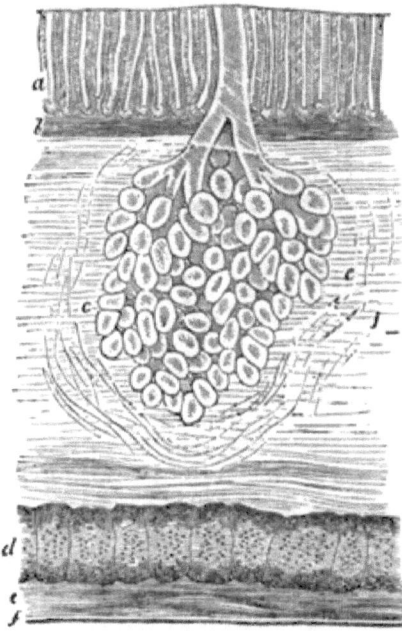

Racemose or duodenal gland, as seen in the vertical section of the duodenum; magnified 40 diameters. *a.* Intestinal tubes. *b.* Muscular stratum of the mucous membrane. *c, c.* Acini of the duodenal gland, which occupies the submucous areolar tissue. *d.* Transverse layer of the muscular coat. *e.* Longitudinal layer of the muscular coat. *f.* Peritoneal tunic of the bowel.

Peyer's glands, or glandulæ agminatæ, occur in the jejunum, but are most numerous in the ileum. They are always found opposite to the attachment of the mesentery, in the form of oblong or rounded patches, as seen in the drawing.

Fig. 105.

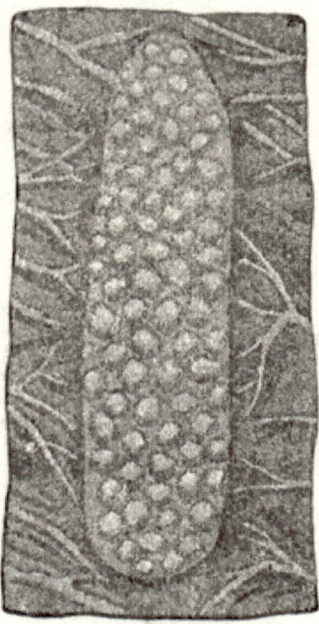

Agminate follicles as seen on a dark ground; magnified about 5 diameters. After Boehm. *a.* General mucous surface of the ileum. *b, b.* Opaque grains corresponding to the several follicles.

Each gland is a closed sac, surrounded by the openings of the intestinal tubes, as will appear when a portion of one of Peyer's patches is more highly magnified.

Fig. 106.

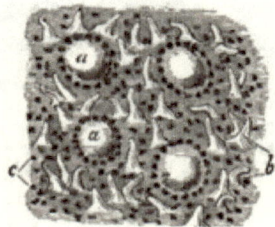

Portion of a cluster of agminate follicles. *a, a.* Follicles encircled by apertures of the intestinal tubes in the form of a ring. *b.* Short and obtuse villi, occupying the intervals of the follicles. *c.* Apertures of intestinal tubes, opening irregularly in these intervals.

In sections, the arrangement of the agminate glands will appear as represented below.

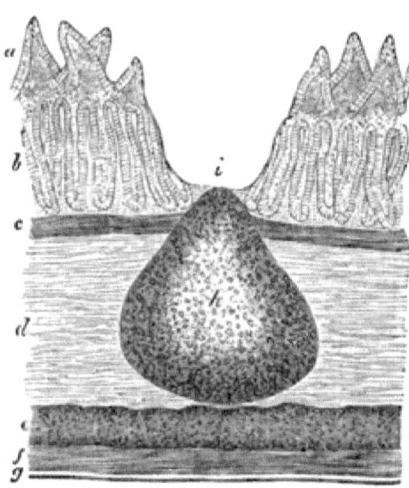

Fig. 107.

Plan of an agminate follicle, as seen by a vertical section; magnified 40 diameters. *a.* Short and conical villi surrounding the follicle. *b.* Intestinal tubes in the same situation. *c.* Muscular stratum of the mucous membrane. *d.* Submucous areolar tissue, in which the follicle is chiefly situated. *e.* Circular layer of the muscular coat. *f.* Longitudinal layer of the same. *g.* Peritoneal coat. *h.* Follicle enclosing nuclear contents. *i.* Apex of the follicle projecting into the cavity of the bowel.

Solitary glands are found scattered through the small intestines in variable number, being most abundant in the jejunum.

When a piece of intestine is held against the light these glands will be seen like grains of mustard-seed. In structure each one so exactly resembles a single agminate follicle, that a separate investigation is unnecessary.

In the large intestines there are no villi. Lieberkühn's follicles, however, are present, and also solitary glands of large size.

Injected preparations necessary for the study of bloodvessels and lacteals may be obtained at such moderate prices, that the student is not advised to attempt the tedious process of making them for himself.

A set of the following will be required :—

Papillæ of tongue.	Brunner's glands.
Œsophageal glands.	Peyer's patches.
Stomach follicles.	Lieberkühn's follicles.
Intestinal villi.	

OF THE PANCREAS.

This gland resembles the salivary glands in structure, being composed of small glandular vesicles, which contain fat-globules, and are lined by epithelium cells. The elements of the gland are united together by connective tissue, in which the vessels, nerves, and ducts are distributed.

For microscopic examination the pancreas of a small animal, as a mouse, rat, or rabbit, is to be selected, and a small portion, if possible a single lobule, cut off with fine scissors, and placed on the slide in a drop of water or serum; the covering glass is to be applied without pressure, and the specimen immediately examined, first under the low, and afterwards under a high power.

If the pancreas of a large animal be employed, it will be necessary to slightly teaze out a small portion, or to press it flat with the covering glass, before putting it under the microscope.

The illustration shows the arrangement of the small vesicles in one of the lobules of the pancreas of a mouse.

Fig. 108.

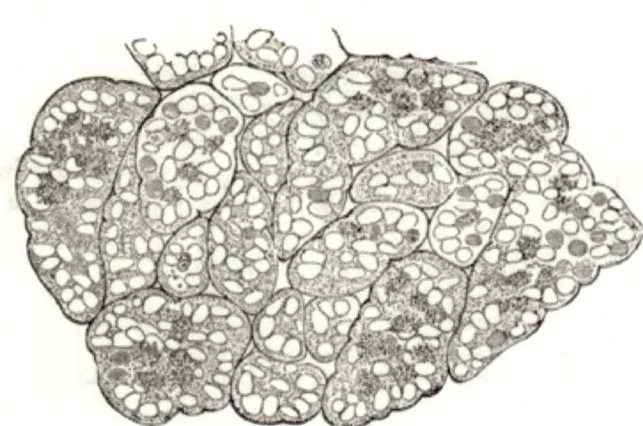

Minute lobule or acinus of the pancreas of a mouse, showing the two forms or stages of the epithelium, and the varied forms and sizes of the ultimate follicles.—Magnified 180 diameters.

The pancreatic fluid may be examined without preparation, a drop being squeezed out from the gland upon the slide, covered with a piece of thin glass, and viewed with a high power. It will be found to consist of a clear colourless liquid, with a few corpuscles, and, occasionally, portions of detached epithelium.

THE LIVER.

The liver is composed of numerous small lobules, polyhedral in figure, varying in size, from half a line to one line in diameter, packed closely together, and connected by fine areolar tissue, intermixed with bloodvessels, nerves, and lymphatics.

The entire gland is invested by two distinct coverings, an outer or serous coat, derived from the peritoneum, and under this a thin capsule of connective tissue.

In order to study the structure of the liver, a small portion of the healthy gland should be obtained from the butcher. The liver of the pig is best, as the lobules in it are most distinct.

An ordinary lens will render the lobules quite evident upon the surface of the specimen without any preparation. They will also be observed in sections, on the cut surface. Each lobule shows the intra-lobular vein in its centre, as indicated in the drawing.

Fig. 109.

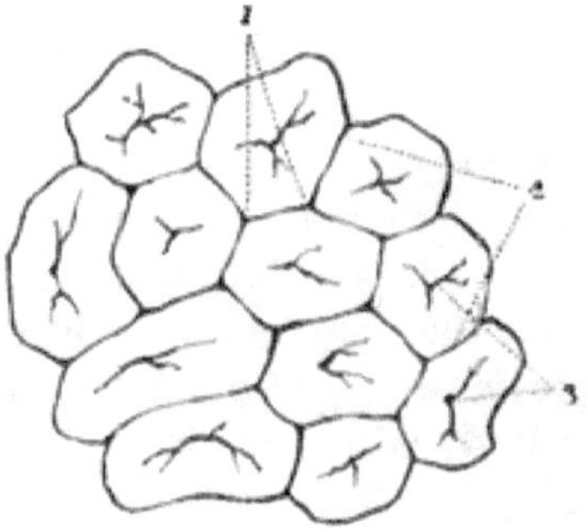

1. Interlobular spaces, containing the larger interlobular branches of the portal vein, hepatic artery, and duct. 2. Interlobular fissures. 3. Intralobular veins formed by minute venules which converge towards the centre of the lobules.

The cells of the hepatic tissue are easily examined in a fresh specimen, by scraping the cut surface with the edge of a knife. Place the matter so collected in a drop of water on the slide, and apply the covering glass in the usual manner. Under a moderately high power the liver-cells will be seen as depicted below.

Fig. 110.

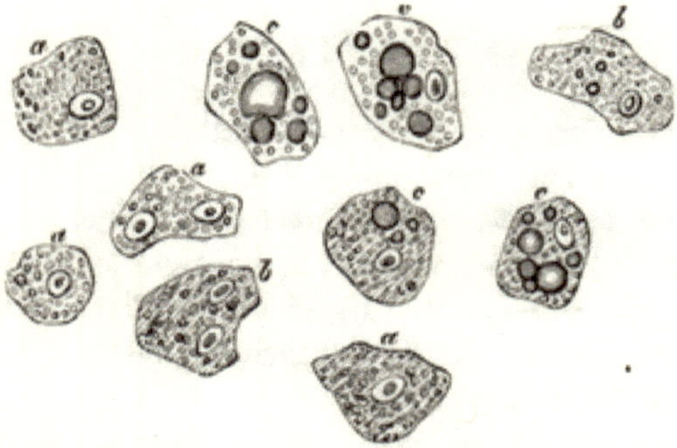

Hepatic cells of man; magnified 400 times. *a*. Normal cells. *b*. With coloured granules. *c*. With fat.

The cells resemble ordinary squamous epithelium; they vary in size from $\frac{1}{1080}$ to $\frac{1}{340}$ of an inch.

In the fœtus two or more large nuclei and nucleoli will be observed, and, in many instances, cells multiplying by the endogenous process; but in the adult the nucleus is often absent. The liver-cells contain, besides the nucleus, granular matter and minute granules of fat. The large fat-globules often seen are the result of disease.

The hepatic cells are enclosed in a delicate membrane, and arranged in radiating lines from the centre to the circumference of the lobule. Sections of the liver may be made by means of Valentin's knife, but they are not very instructive unless the gland has been previously injected. Such preparations are indispensable for the study of the vessels.

The next drawing represents some of the tubes filled with hepatic cells. The liver of the pig is best for making this preparation. A small portion should be carefully teazed out, covered in the usual way, and examined with the high power.

Fig. 111.

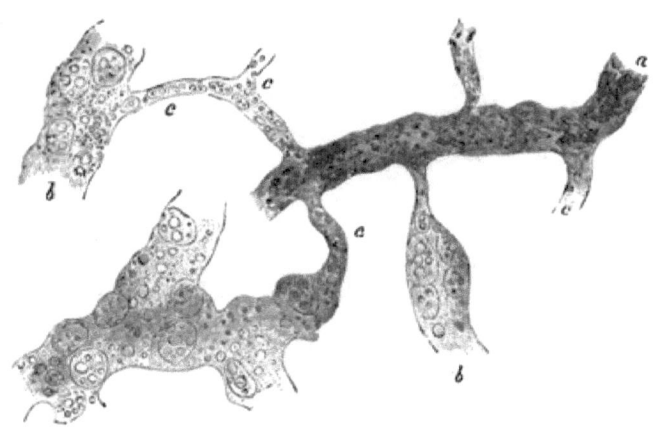

Termination of a small interlobular duct in the pig's liver, and communication of its smallest branches with the network of tubes containing liver-cells.

Bile does not generally contain any microscopic elements of importance. Epithelial cells from the gall-bladder will usually be observed, with small masses of colouring matter, and, occasionally, plates of cholesterine.

THE THYROID GLAND.

MICROSCOPIC examination of the thyroid body demonstrates the existence of numerous vesicles or loculi lined with epithelium, and distributed through a fibrous stroma.

Preparation.—Select specimens of young subjects in preference—those from birds, dogs, and rabbits answer very well.

Sections may be made by the double knife, or small portions may be teazed out in a little water on the slide, covered with the thin glass, and viewed with the high power. Several preparations may be necessary before the characteristic appearance indicated in the drawing is seen.

Fig. 112.

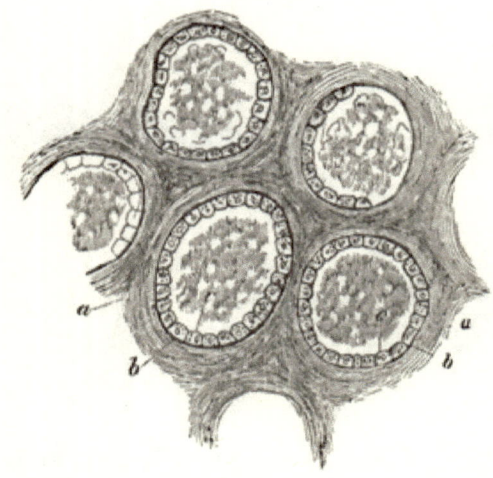

Some gland-vesicles from the thyroid gland of a child; magnified 250 times. *a.* Areolar tissue between the same. *b.* Membrane of the vesicles. *c.* Their epithelium.

Injected specimens will be required, in order to study the disposition of the vessels, which are best seen in sections near the surface. In many specimens no defined epithelium will be distinguished; but the loculi or vesicles will be seen to be filled with a fluid containing numerous dark granules, probably the result of a post-mortem change.

THE THYMUS GLAND.

The thymus gland is formed of a number of lobules united by connective tissue, and attached to a common stem in the centre of which a canal exists. The arrangement of the parts is shown in the accompanying sketch of the thymus of the calf. In man the central canal is of an irregular form.

The duct generally runs in an irregular spiral course through the middle of the gland; upon its centre there are to be observed a number of openings leading into the several lobules.

Fig. 113.

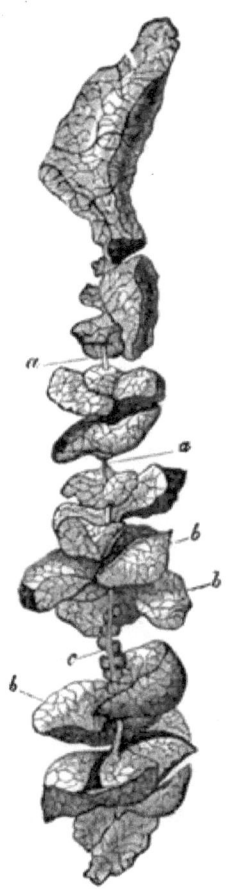

A piece of the thymus of the calf, spread out. *a.* Chief canal. *b.* Lobules. *c.* Isolated gland-granules, seated upon the principal canal.—Natural size.

Preparations of the gland for the purpose of examining its intimate structure may be made by boiling a portion, and afterwards steeping it in alcohol. Hardened in this manner it may cut into fine sections.

Each lobule will be found to have a cavity in the centre, and to be separable into a number of smaller subdivisions, and again into small round bodies or gland granules (acini of authors).

The woodcut represents a single lobule with its subdivisions, central cavity, and acini. The drawing is taken from an injected specimen.

Fig. 114.

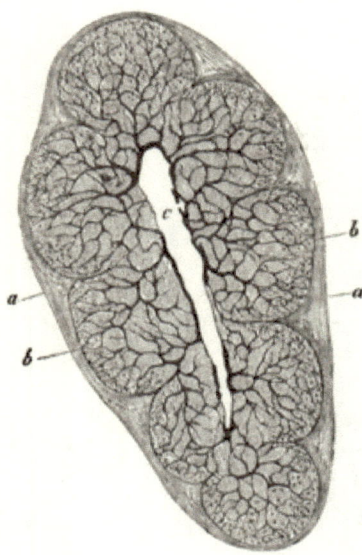

Transverse section through the apex of an injected lobule of the thymus of a child; magnified 30 times. *a.* Envelope of the lobule. *b.* Membrane of the acini. *c.* Cavity of the lobule, from which the large vessels ramify into the acini, and partly terminate with loops upon the surface of them.

The elements of the gland can be examined by teazing out a fragment in a drop of water on a glass slide. The covering is to be applied in the usual way, and the specimen viewed under a high power. Numerous cells will be seen, also free nuclei, fat-granules, and the fibres of connective tissue.

The secretion contained in the central canal of the thymus is a clear albuminous fluid in which are distributed cells, free nuclei, and sometimes peculiar concentric bodies whose nature is not apparent. Considerable difficulty attends the minute investigation of the thymus, but the fœtal organ will be found much the easiest to prepare.

THE SPLEEN.

THE spleen has two investing membranes; one derived from the peritoneum, and a second (the *tunica albuginea*) of fibrous tissue intermixed with elastic fibre and in some animals with smooth muscular fibre.

On making a section of the spleen a peculiar arrangement of its fibrous structure is seen, particularly in a specimen that has been well washed.

The white fibres interlace in all directions, forming the trabeculæ of the spleen, as shown in the illustration.

Fig. 115.

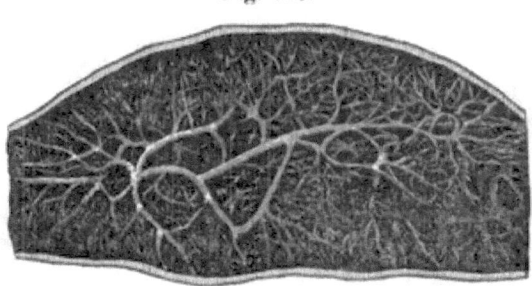

Transverse section through the middle of the spleen of the ox, washed out in order to show the trabeculæ and their arrangement. Natural size.

A beautiful section of the trabecular tissue can be readily made by cutting a slice an inch thick from the spleen of an ox, and then squeezing it in tepid water like a piece of sponge. In a few minutes the pulp will be broken up and floated away, leaving nothing but the network of fibres, which may be said to form the skeleton of the spleen.

The soft red substance, the splenic pulp, which fills the spaces in the trabecular tissue, may be squeezed out, or scraped from the cut surface with a scalpel. A little of it should be placed on the slide in a drop of serum or glycerine, covered in the usual way, and examined with a high power.

The elements of the spleen pulp are not always the same, but illustrations are given of all that are likely to be met with: viz.,

128 HISTOLOGY.

cells with large nuclei, blood discs, cells enclosing blood corpuscules, and spindle-shaped cells from the splenic vessels.

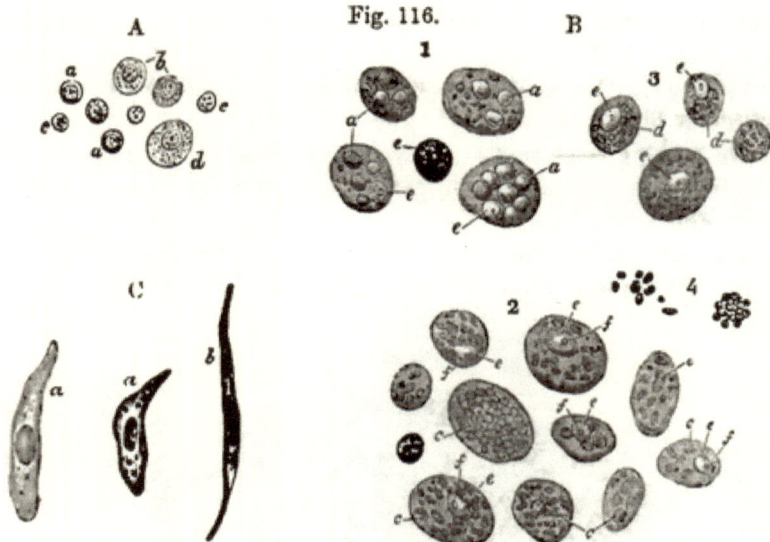

Fig. 116.

A. Parenchyma-cells from the spleen of the ox, magnified 350 diameters. *a.* Smaller cells. *b.* Cells of medium size. *c.* Free nuclei. *d.* Largest cells.
B. Cells containing blood corpuscles from the spleen of the rabbit, magnified 350 diameters. 1. Cells with one, three, four, and seven unchanged blood corpuscles. 2. Cells with blood corpuscles undergoing dissolution, and coloured in different shades of brown or yellow (coloured granule cells). 3. Cells with destroyed and decolorised blood globules, larger or smaller, and with or without granules. 4. Blood globules altered in colour, diminished or destroyed, either single or aggregated, in small lumps.
In 1, 2, and 3, the following letters signify alike :—
a. More or less unchanged blood globules. *c.* Coloured granules arising from a diminution or destruction and alteration of colour in blood corpuscles. *d.* Colourless granules produced by the discolouration of *c*. *e.* Nuclei of the cells containing blood corpuscles and their metamorphoses. *f.* Nucleoli of these nuclei.
C. Epithelial cells from the human splenic artery. *a.* Shorter cells. *b.* A somewhat longer cell.

Imbedded in the red pulp of the spleen will be found numerous little white bodies, which are the Malpighian corpuscles connected with the minute arteries.

In order to see these bodies the spleen of the ox or sheep should be examined. Make a section through the organ, and look carefully into the red substance, if necessary using a common lens to

THE SPLEEN—THE KIDNEYS.

assist the search. The small rounded bodies being discovered, one of them may be insulated with the needles, and placed in a drop of water on a glass slide, the covering glass being applied with sufficient pressure to slightly flatten the corpuscle. The observation should be made with the high power, and if the preparation has been properly made, the wall of the corpuscle and its contents will be discerned as indicated in the drawing, which shews a single Malpighian body highly magnified. Generally some of the elements of the pulp with minute trabeculæ and small arterial twigs will be seen in addition.

Fig. 117.

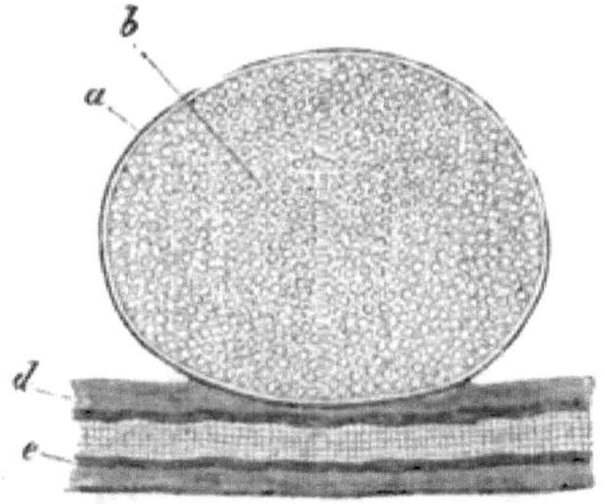

A Malpighian corpuscle, from the spleen of the ox; magnified 150 times. *a.* Wall of the corpuscle. *b.* Contents. *c.* Wall of the artery, upon which it is seated. *d.* Sheath of the same.

The distribution of bloodvessels can only be studied in injected specimens.

OF THE KIDNEYS.

As it is occasionally useful to be able to distinguish one kidney from the other, the following characteristics should be noticed.

1. The largest ends are uppermost; the anterior surfaces are convex, and the posterior flat.
2. The right kidney is rather shorter and wider than the left.
3. At the notch or hilus the vessels enter and emerge in the

following order: the vein is in front; the artery next; the ureter behind and lowest down.

In man, pig, sheep, deer, and horse, the kidneys are non-lobulated.

In the bear, dolphin, and snake they are lobulated, as also, though to a less degree, in the ox.

The human fœtal kidney is lobulated, but the lobes disappear about the time of birth; some specimens, however, retain the fœtal condition during life.

Preparation.—Obtain from the butcher a healthy and perfectly fresh sheep's kidney, and cut it lengthwise in the same way that the cook does previous to dressing it; this section will expose the outer or cortical substance and the inner or medullary structure, with the pelvis, the ureter, the mamillary cones, and the calices, as shown in the drawing.

Fig. 118.

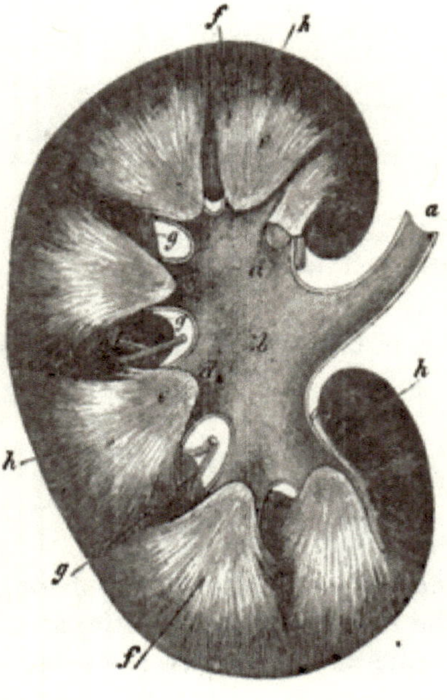

A section through the middle of the kidney of a child. *a.* Ureter. *b.* Pelvis. *c.* Calyces. *d.* Papillæ. *e, f.* Malpighian pyramids. *g.* Septa Bertini. *h.* Outer part of the cortical substance.

The medullary portion of the kidney is the easiest to prepare for microscopic examination, and consequently should be taken first.

With a scalpel scrape sharply across the mamillary bodies, and place the matter so obtained in a drop of water on the slide, apply the covering glass gently, and examine with the high power. If too much tissue is in the field there will be nothing definite seen, in which case remove the specimen, add a little more water, and, if necessary, separate the particles with the needles.

A good preparation will show, as in the drawing, the uriniferous tubes, some lined with epithelium, and others nearly or quite empty; only the delicate membrane forming the walls of the tubes with a few scattered cells being visible. Very few blood-vessels will be observed in the preparations of the mamillary portion of the kidney.

Fig. 119.

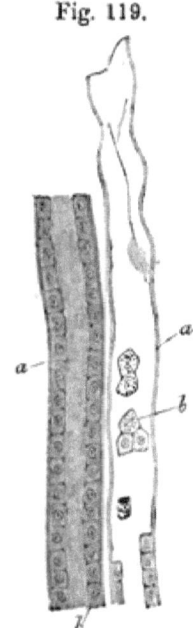

Two straight urinary tubules of man, the one with complete epithelium, the other nearly empty. *a.* Membrana propria. *b.* Epithelium.

The next step will be to scrape the cortical part lightly with the scalpel in a direction towards the pelvis of the kidney. The

132   HISTOLOGY.

matter so obtained must then be distributed through a little water on the centre of the glass slide, and covered in the usual way.

The preparation, if successful, should show under a low power the Malpighian tufts with tortuous uriniferous tubes proceeding from them, but the Malpighian bodies will not be seen distinctly excepting in injected specimens.

It will be noticed in the course of numerous examinations that all the uriniferous tubes do not arise from Malpighian tufts, but that many of them commence in cæcal extremities.

The most perfect observation of the structure of the Malpighian tuft and the uriniferous tube can only be made in successful preparations of the injected gland.

The next drawing shows the relations of the various parts very distinctly.

Fig. 120.

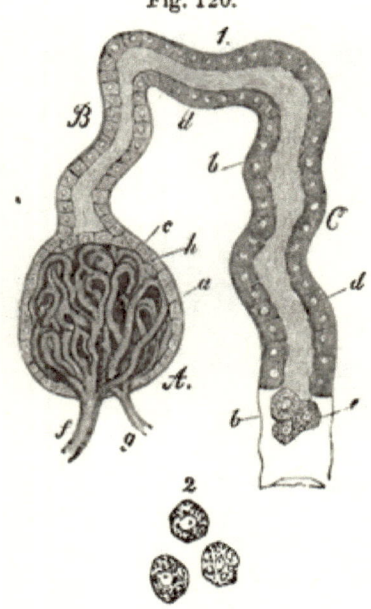

1. A human Malpighian corpuscle, *A*, with the urinary tubule, *B C*, arising from it; magnified 300 times. Half-diagrammatical figure. *a.* Envelope of the Malpighian corpuscle, continuing into *b*, the *membrana propria* of the convoluted uriniferous tubules. *c.* Epithelium of the Malpighian corpuscles. *d.* Epithelium of the uriniferous tubule. *e.* Detached epithelial cells. *f.* Vas afferens. *g.* Vas efferens. *h.* Glomerulus Malpighianus. 2. Three epithelial cells, from convoluted tubules. Magnified 350 times. One with fat-drops.

Transverse sections of the kidney are very easily made with the double knife either through the cortical or medullary portions. The sections are best made under water, and thence floated on to the glass slide, covered in the usual manner, and viewed first with a low power. The arrangement of the tubes in the stroma will be very beautifully shown, as in the drawing, if the sections have been made with ordinary care; it is not necessary that they should be extremely thin.

Fig. 121.

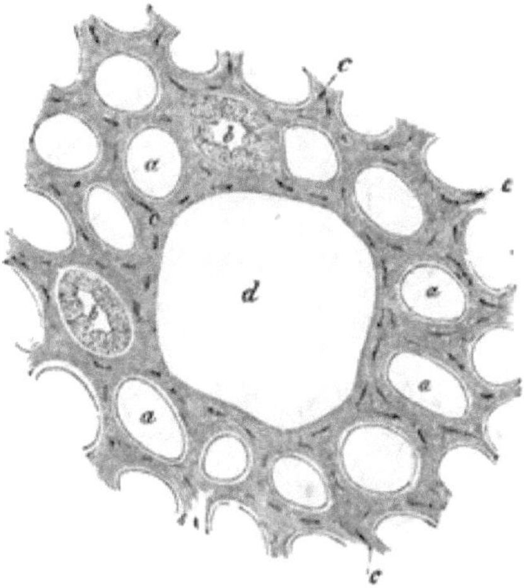

Transverse section through some straight tubules of the cortex of the human kidney; magnified 350 times. *a.* Transverse sections of urinary tubules, of which the *membrana propria* is alone preserved. *b.* The same where the epithelium is still present. *c.* Stroma of areolar tissue, with elongated nuclei. *d.* Space, which has contained a Malpighian corpuscle.

SUPRA-RENAL CAPSULES.

These bodies may be investigated by the student with comparative ease after they have been hardened in chromic acid, but in their fresh condition it is exceedingly difficult to make the sections sufficiently thin, although it may be done by the aid of the double knife, the structure being fastened upon a piece of cork or wood, and held under water while the section is made.

134   HISTOLOGY.

The hardened capsules, on the other hand, can be cut with the razor, and for the purpose of observing the relations of the various parts, one section should be made through the entire length of the organ, and another transversely; the specimens should next be placed in equal parts of glycerine and water on the slide, covered in the usual way, and examined first with the low power.

If the preparations are successful, the parts represented in the drawing will be distinguished.

Fig. 122.

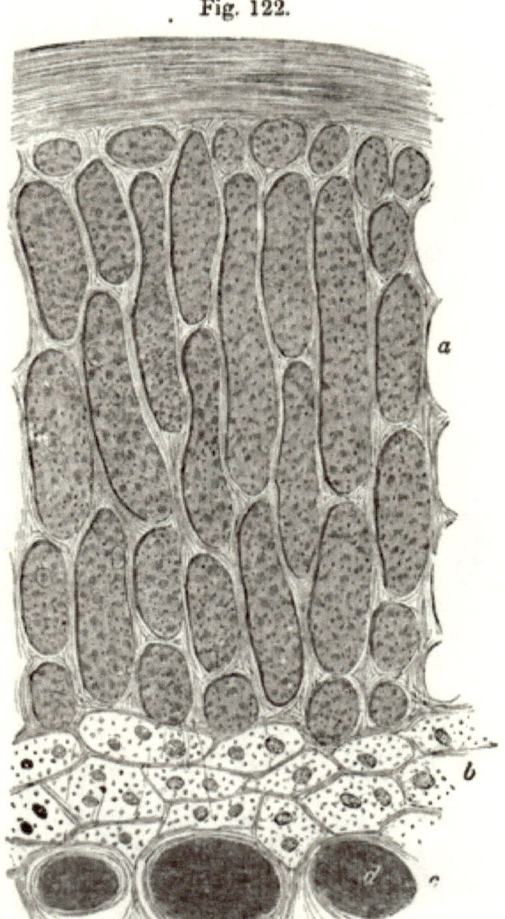

Transverse section of a human supra-renal capsule. *a.* Cortical, *b.* Medullary substance. *c.* Sinuses.

The supra-renal capsule when cut across is seen to be composed of two differently coloured portions. The external or cortical substance is of an orange colour, becoming yellow as it reaches the internal or medullary part, which is, when filled with blood, dark red; when washed, of a slate colour. In this medullary substance several round or oval openings of venous sinuses may be seen by the aid of a lens. Under the low power the cortical substance is seen to consist of cells arranged in rows, appearing like dark yellow columns, placed perpendicularly to the surface, and terminating abruptly at the margin of the medullary substance.

With the high power the cells composing the columnar masses will be found to contain a nucleus, which is very distinct in preparations that have been coloured in the carmine solution; besides the nucleus there are numerous granules and oil-globules.

The medullary substance of the capsule is composed of a fine reticulated fibrous structure, in the meshes of which lie numerous large pale cells with round nuclei, altogether not unlike ganglion corpuscles.

Vessels and nerves are freely distributed to the renal bodies, but for their observation injected preparations are indispensable.

LUNG.

The lungs are porous spongy structures, having a specific gravity of 500 (water being 1000). When fully distended their specific gravity is 126; when entirely free from air, it is 1056.

Colouring matter is distributed in the interstitial cellular tissue, and sometimes in the air-cells.

Preparations of Lung for the Microscope.—Inflate the lungs of a small animal, tie the root of the trachea, and dry the organs slowly; or otherwise inject them from the trachea with melted size, and, when cold, cut sections with a razor. As a rule, however, the inflated lung is best; it may be easily cut into fine slices, and if these are steeped for twenty-four hours in the carmine solution before being examined, the walls of the air-cells will be rendered beautifully distinct.

To make sections of fresh lung, a portion must be fixed upon the leaded cork, and cut under water with the Valentin's knife. The section may then be floated on to the slide, and covered in the usual way.

When the dried lung is employed, a thin section should be cut with the razor, and placed in a little water on the slide; the covering glass being then applied, the object is ready for examination, and should be viewed with the low power first, and afterwards with the high. If the sections have been carefully made the air-vesicles lined with epithelium will be seen as depicted.

Fig. 123.

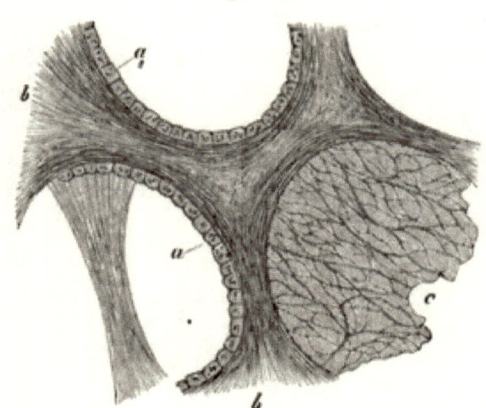

A human pulmonary vesicle, with the parts adjoining; magnified 350 times. *a.* Epithelium. *b.* Elastic trabeculæ. *c.* More delicate walls between the trabeculæ, with fine elastic fibres.

Externally the lungs are covered by serous membrane (pleura), beneath it is the subserous (areolar) tissue, containing many elastic fibres, and some smooth muscular fibre.

In the lungs of lion, seal, and leopard, the subserous tissue forms a strong elastic membrane.

The substance of the lung consists of small lobules attached to the ramifications of the air tubes; they are of various sizes, polyhedral in figure, with flattened sides.

On the surface they are pyramidal, with lozenge-shaped bases, turned outwards.

The lobules are adherent but quite distinct from each other, and in the fœtal lung easily separated.

The illustration represents two lobules, with terminal bronchial tubes.

Fig. 124.

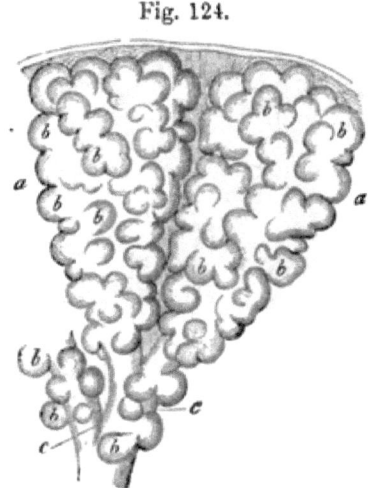

Two small pulmonary lobules. *a, a.* With the air-cells, *b, b,* and the finest bronchial branches, *c, c,* on which air-cells are likewise seated, of a newly born infant.—Magnified 25 times. Half-diagrammatical figure.

Each lobule is composed of a number of air-cells, honeycomb-shaped, or quadrilateral areolar cavities, with very thin walls.

The air cavities are lined by squamous epithelium, and do not open into each other.

To study the vessels of the lungs injected specimens will be necessary. These are so cheaply obtained that the student will hardly incur the trouble and loss of time necessary for their preparation. The most instructive specimens are those made by transparent injection, although very beautiful preparations are obtained by the opaque method, which is very successfully applied to the lungs of reptiles.

The distribution of the bloodvessels in relation to the pulmonary vesicles is well shown in the accompanying drawing.

Fig. 125.

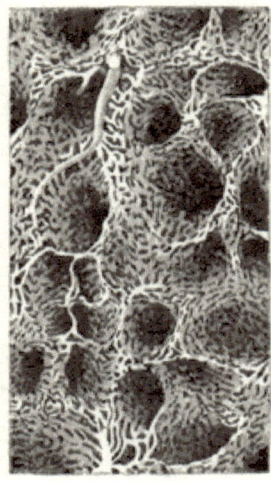

Capillary network of the pulmonary vesicles of man.—Magnified 60 times.

ARTERIES, VEINS, AND CAPILLARIES.

Arteries and veins are best examined in sections of dried preparations. A large vessel, the aorta or vena cava of an ox for example, is to be selected, and when sufficiently dried to bear the pressure of the razor, fine sections are to be made longitudinally as regards the vessel, which will be transversely to its circular fibres. These sections may be placed in a little water on the slide, and covered after they have become sufficiently softened, or, what is better, they may be steeped in carmine solution, which renders the several layers much more distinct.

Small vessels can be dried by spreading them out on a piece of card. The sections are to be made through both card and vessel; the separation of the two takes place as soon as the preparation is moistened on the slide.

ARTERIES, VEINS, AND CAPILLARIES. 139

Under the low power a longitudinal section of a large artery shows the various coats of which the vessel is composed very distinctly.

Fig. 126.

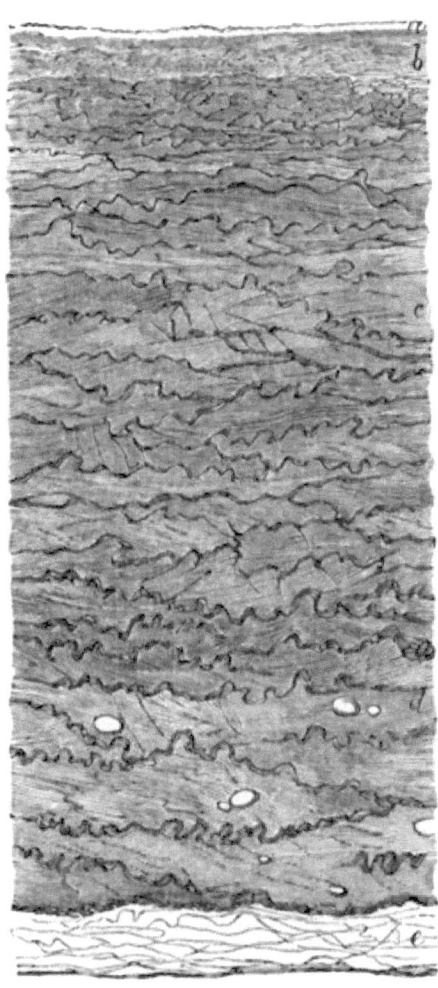

Longitudinal section of aorta of horse. *a.* Epithelial layer. *b.* Longitudinal layer of fine elastic fibres. *c.* Longitudinal layer of yellow fibre. *d.* Circular layer of yellow elastic fibres. *e.* External coat of areolar tissue.

1. An external investure of areolar tissue.

2. A layer of coarse yellow elastic fibres, with some plain muscular fibres, arranged in a circular or oblique direction. The elastic fibres of this layer branch in a penniform manner, as will be seen when a small portion of the tunic is teazed out and examined with a high power. Between these penniform fibres the smooth muscular fibres are found.

3. A layer of coarse yellow fibres running in a longitudinal direction, and interlacing closely with each other.

4. A thin layer of fine elastic fibres, also taking a longitudinal course. This tunic may be readily stripped off the interior of an artery, by first cutting it across very lightly, and then taking the edge of the cut membrane in the forceps and gently peeling it away. Under a high power the fine elastic fibres will be apparent. Sometimes the membrane is pierced with numerous apertures, constituting the fenestrated membrane of Henle.

5. A layer of oval or fusiform epithelial cells, which should be scraped from the interior of an artery of an animal recently killed, as they are separated soon after death.

The fine elastic layer with the epithelium represents the serous lining membrane of the older authors.

In the illustration, taken from a longitudinal section of the aorta of a horse, the different tunics are shown in their relation to each other.

Veins.—The coats of veins are thinner than those of arteries, but correspond to them in structure and arrangement.

Thus we observe an external coat of areolar tissue; next a circular arrangement of course yellow fibres, with plain muscular fibres and white fibrous tissue, corresponding to the circular fibrous coat of arteries, but much less developed: then the two layers of yellow fibres arranged longitudinally, the outer layer being coarse and the inner fine in texture, and having upon its surface the single layer of epithelial cells.

The majority of veins possess semilunar processes, or valves projecting into their interior. These valves are composed of the same tissue as the fine fibrous tunic, and are covered by a layer of epithelium.

Capillaries.—Intermediate between arteries and veins, there are exceedingly fine hair-like tubes, termed capillaries, in which arteries terminate, and from which veins arise, excepting in a very few instances where arteries communicate directly with veins, as in the cavernous structure of the penis.

Capillaries are composed of a fine homogeneous membrane, with here and there oval nuclei. In some of the larger tubes a distinct arrangement of circular fibres may be seen, resembling smooth muscular fibre, but differing from it in the absence of any trace of nuclei.

In some of the organs of the body, the liver for example, it is impossible to distinguish distinct walls of the capillary vessels, in the mass of epithelial cells in which they are imbedded.

The drawing represents some of the capillaries of the human brain.

Fig. 127.

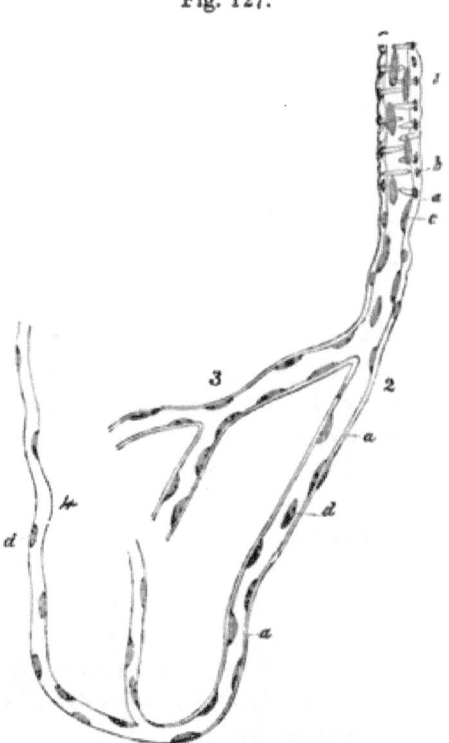

Finest vessels of the arterial side. 1. Smallest artery. 2. Transition vessel. 3. Coarser capillaries. 4. Finer capillaries. *a.* Structureless membrane, still with some nuclei, representative of the tunica adventitia. *b.* Nuclei of the muscular fibre-cells. *c.* Nuclei within the small artery, perhaps appertaining to an epithelium. *d.* Nuclei of the capillaries of the transition vessels. From the human brain.—Magnified 300 times.

Lymphatic vessels possess still more delicate coats than veins, to which they bear a general resemblance in the arrangement of their elements, and in the existence of valves in their interior.

The external investment is composed of areolar tissue; next there is a layer of circular fibres with fibres like those of unstriped muscle, having elongated nuclei; then a longitudinal coat of fine fibres, like those of connective tissue; and upon this a single layer of epithelial cells.

OF THE NERVOUS SYSTEM.

The study of the tissues belonging to this system will be much facilitated by a judicious selection of specimens.

In order to obtain a view of an entire nerve, one of the small cutaneous dorsal branches should be taken from the frog, by dividing the skin of the back carefully along the centre of the spinal column, avoiding to cut the roots of the nerves.

On reflecting back the integument the small superficial nerves will be exposed to view, like so many fine threads, extending from the muscles to the skin. A portion, about one-eighth of an inch, of one of the smallest of these is to be cut off with fine scissors, placed in a drop of water on the slide, and gently drawn into a straight line by the aid of the needles; the thin glass cover is then to be applied, and the specimen at once examined with the low power.

The object is an exceedingly beautiful one if well prepared, consisting of a sheath of areolar tissue, with a bundle of nerve fibres in its centre, and frequently a small bloodvessel enclosed in the same sheath, as shown in the illustration.

The addition of a drop of acetic acid will render the sheath much more transparent, and, after a few seconds, bring into view the 'connective tissue corpuscles.'

The nerve will generally appear crooked, as in the drawing; this arises from the contraction of the fibrous sheath, which had previously been stretched, but after being cut shrinks considerably and causes the appearance observed.

NERVOUS FIBRILLÆ. 143

Fig. 128.

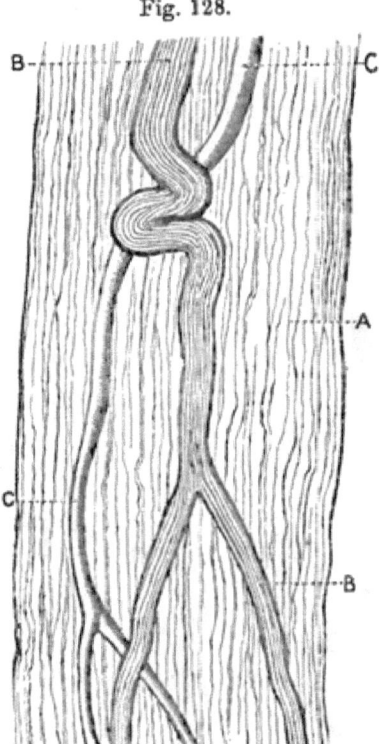

Branch of nerve from the dorsal region of a frog. A. Areolar tissue. B. Bundle of nerve fibres assuming a serpentine course and dividing into two branches before leaving the common sheath. C. Bloodvessel also branching.

Examination of Nervous Fibrillæ.—After a general observation of the structure of an entire nerve, the student may proceed to examine the fibres of which it is composed. For this purpose it is best to select the largest nerve in the frog's body, the sciatic, which will readily be found as a thick white cord, lying between the muscles at the back of the thigh.

A piece of this nerve, about one-eighth of an inch long, is to be cut off and placed on the slide in a little water, or, what is much better, serum or white of egg, and carefully teazed out in a longitudinal direction, that the minute fibres may be separated from each other without being broken; the covering glass should then be applied very lightly, and the specimen examined with a high power.

The ultimate fibres of the nerve will be recognised by their peculiar transparency, appearing like fine threads of spun glass. After a few minutes this pellucid aspect is lost, and the fibres are seen to consist of the three parts shown at 3 in the illustration, which represents the nerve fibres of several animals.

Fig. 129.

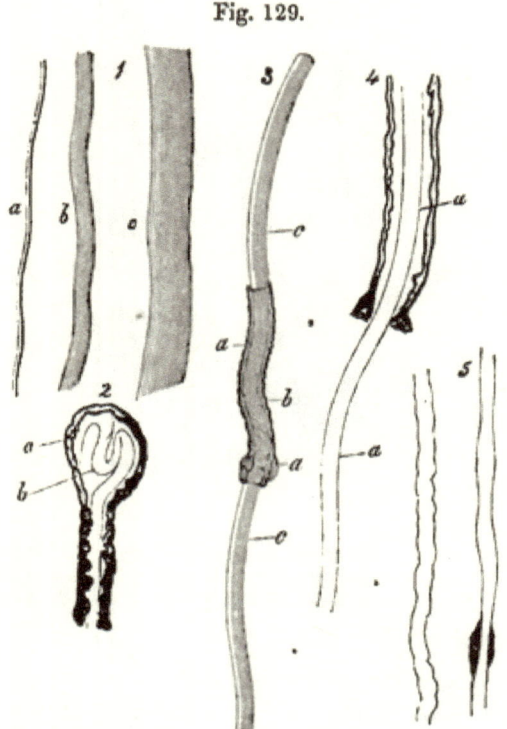

Nerve fibres; magnified 350 times. 1. Of the dog and rabbit, in the natural condition. *a.* Fine, *b.* middling thick. *c.* Thick fibre, from the peripheral nerves. 2. Of the frog, with the addition of serum. *a.* Drop forced out by pressure. *b.* Axis cylinder in the same, continuing into the tube. 3. Of the fresh spinal cord of man, with the addition of serum. *a.* Investment. *b.* Medullary sheath, with double contours. *c.* Axis cylinder. 4. Double contoured fibre of the human fourth ventricle, the axis cylinder (*a*) projecting and visible in the fibre. 5. Two isolated axis cylinders from the cord, the one undulated, the other unequally thick, with adherent medulla.

In size the fibrillæ vary, even in the same nerve, from $\frac{1}{12000}$

to $\frac{1}{1500}$ of an inch in diameter, according as they are nearer to the root or the terminal branches of the nerve.

The sheath (neurolemma) is an exceedingly delicate membrane, quite homogeneous and perfectly transparent, and does not come properly into view until the white substance has undergone coagulation, and become irregular in its outline. Then the fine sheath is seen stretching across the hollows formed in the tortuosities of the white substance.

The white or medullary substance after coagulation refracts the light, so as to give rise to the appearance of a dark shaded border on each side the nerve tube. The border being very much darker than the rest of the tube, slightly brown or yellow in tint, and bounded by two parallel lines, is usually very distinct and constitutes the 'double contours' of the white substance. These double contours are only seen in fibres of a certain size, and are due to the shortening of the sheath from its own resiliency, and the narrowing of the white substance by its coagulation producing less room lengthways than sideways. This is partly proved by the fact that in fine nerve fibres, which are only dilated at intervals, the double contour is only seen in the dilated portions of the tube.

The Axis Cylinder, or primitive band, which fills up the entire centre of the nerve tube, is more tenacious than the white substance, and is often seen projecting beyond it, sometimes spreading out like a drop at the end of a broken fibre.

Occasionally the axis cylinder appears to be striated longitudinally, and has been seen to split into fine filaments.

Like all solid cylinders, it has a double outline, which is merely the result of the sides not being in focus at the same time as the centre.

Gelatinous or Grey Fibres.—Hitherto we have been speaking of the white nerves belonging to the cerebro-spinal system; we have now to examine those more particularly associated with the sympathetic system, and distinguished as the grey or organic fibres.

Grey fibres are always found in combination with white or tubular fibres, and exist in small proportion in the cerebro-spinal nerves, but in the sympathetic they occur in great numbers.

For the purpose of examination it will be necessary to obtain a piece of the sympathetic from the neck of a rabbit or cat, in which animals the nerve is found lying quite free alongside the pneumogastric, and is readily distinguished by the ganglia attached to it.

The method of preparation has already been explained in reference to the white fibres. A small portion of the nerve, after being teazed out and covered in the usual way, will show, under the microscope, numerous fibres, which are as a rule smaller than tubular fibres, and present the appearance of fine flattened homogeneous bands, some of them having corpuscles upon them at irregular intervals, as seen in the illustration.

Fig. 130.

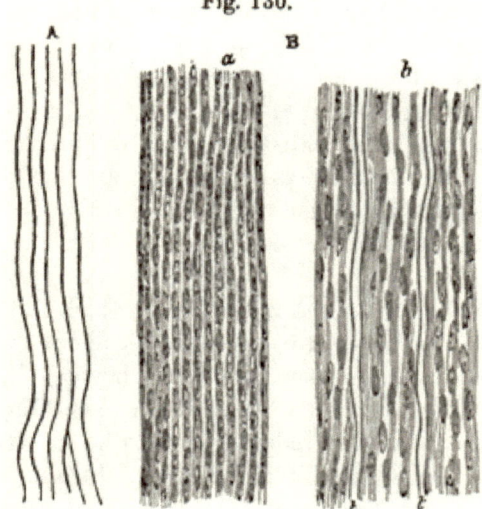

A. Nerve fibres from the brachial plexus of a calf 18 inches long. B. *a.* Nerve fibres from the sympathetic cord in the thorax. *b.* The same treated with dilute solution of soda, showing the presence of tubular nerve fibres, *t, t,* similar to those in the cerebro-spinal nerves.

The gelatinous fibres are always mixed up with a considerable quantity of areolar tissue, which is not unfrequently mistaken for nerve fibres. This error may be avoided by allowing a drop of acetic acid to flow under the covering glass. In a few seconds the fibrous tissue will swell up and become indistinct, while the nerve fibres remain unaltered, except that their corpuscles are brought more distinctly into view.

Besides nerve fibres there are nerve cells or vesicles, which are found in the nerve centres. In structure they consist of a cell wall filled with granular matter, and contain both nuclei and nucleoli; in figure they are generally spheroidal, but sometimes they are angular, oblong, or irregular, and many of them are caudate, especially those from the grey matter of the brain and spinal cord. These caudate vesicles are named uni-, bi-, and multi-polar, according to the number of processes they possess.

Nerve cells also occur in the ganglia, and are very much larger than those found in the brain, varying from $\frac{1}{3000}$ to $\frac{1}{300}$ of an inch in diameter.

Ganglia cells are surrounded by groups of granular corpuscles, in which nuclei are imbedded.

In order to examine these vesicles, a portion of a ganglion from the sympathetic must be well teazed out in a little water on the glass slide, covered in the usual manner, and viewed with a high power.

A drop of acetic acid placed at the edge of the covering glass will render the capsule with its nuclei quite distinct, as shown in the illustration.

Fig. 131.

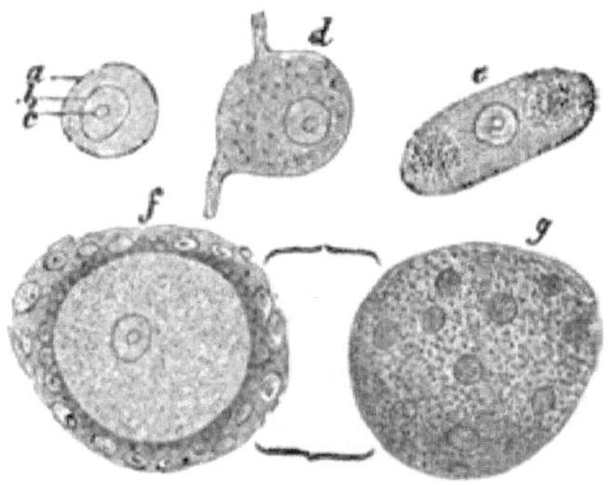

Nerve vesicles from the Gasserian ganglion of the human subject. *a.* A globular vesicle with defined border. *b.* Its nucleus. *c.* Its nucleolus. *d.* Caudate vesicle. *e.* Elongated vesicle with two groups of pigment particles. *f.* Vesicle surrounded by its sheath or capsule of nucleated particles. *g.* The same, the sheath only being in focus.

Caudate cells may be obtained for examination by gently

scraping a small portion from the grey matter of the brain or spinal cord, and lightly teazing it out in a drop of water on the slide. It will, however, happen in spite of all precautions, that many of the processes of the vesicles will be broken off during the preparation, and for this reason a previous hardening of the brain in chromic acid is to be recommended.

Fig. 132.

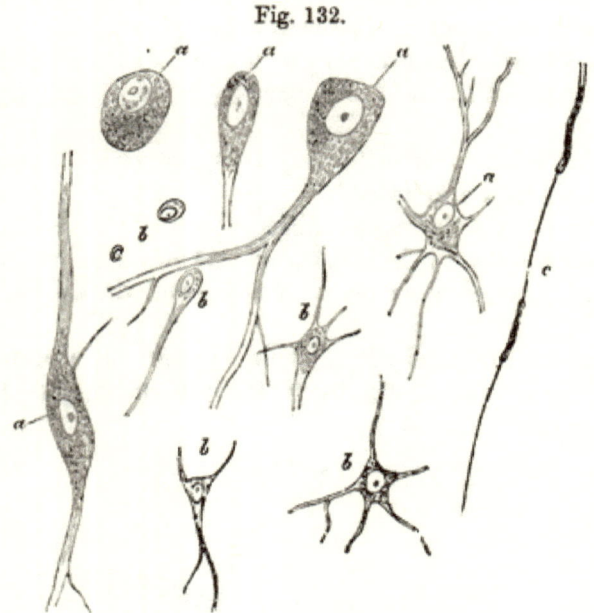

From the inner parts of the grey layer of the convolutions of the human brain; magnified 350 times. Nerve-cells, *a*, larger, *b*, smaller. *c*. Nerve-fibres with axis-cylinder.

Pacinian Bodies are small oval clear ice-like structures, varying from $\frac{1}{15}$ to $\frac{1}{10}$ of an inch in length, and $\frac{1}{25}$ to $\frac{1}{20}$ of an inch in breadth, situated on the terminal twigs of certain nerves, especially those of the hands and feet.

In some of the lower animals they are very highly developed; and for examination the mesentery of the cat will afford good specimens.

If a portion of the mesentery be spread out on the glass slide, the Pacinian bodies will easily be seen with the unaided eye, lying between the transparent folds of the membrane, like small oval seeds.

PACINIAN BODIES.

One or two of these seed-like bodies are to be left on the slide. A drop of water should be added, and the covering glass applied without any pressure.

If the object has been properly arranged it should appear, under the low power, as an oval body composed of numerous lamellæ, with a cavity in the centre, in which lies the termination of of the nerve, and often by the side of it a small arterial twig.

Under a high power the lines of the lamellæ are more distinct, and a drop of acetic acid applied to the edge of the covering glass will render a number of nuclei apparent upon them.

Fig. 133.

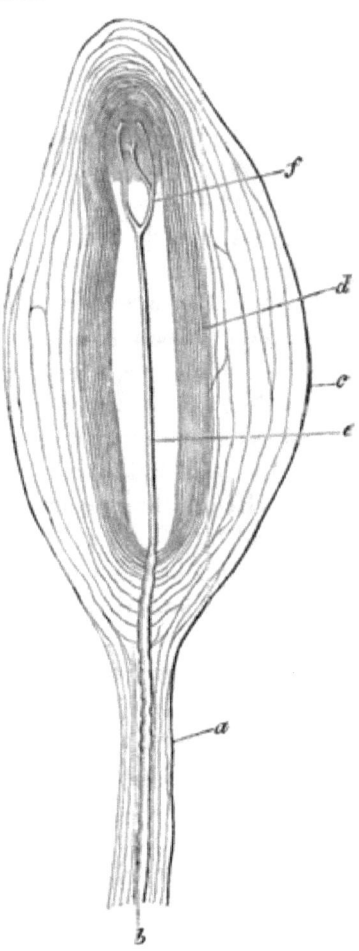

A human Pacinian body; magnified 350 times. *a.* Pedicle of the same. *b.* Nerve-fibre in it. *c.* Outer, *d.* inner layers of the envelope. *e.* Pale nerve-fibre in the central cavity. *f.* Division and termination of the same.

The Brain and Spinal Cord.—In order to examine the structure of these parts, it is necessary to harden portions, by immersion in chromic acid or alcohol, until they are sufficiently firm to be cut with a razor. The sections are afterwards to be rendered transparent by being covered with glycerine.

Mr. Lockhart Clarke's method of making preparations of nerve structure has been found so effective, that the process is given in detail:—

Portions of brain, medulla, or spinal cord, are to be cut into pieces, as small as convenient for making the necessary sections subsequently, and steeped in a solution of one part of crystallised chromic acid to 200 parts of water for two or three weeks, and then kept in a solution of about 1 part of bichromate of potash in 100 to 200 parts of water. If for rodents, birds, reptiles, and fishes, 1 part to 600 parts of water, gradually increasing the strength at the end of a week.

Sections may then be made by means of a knife or razor dipped in spirit. The slices are to be first placed in spirit for a few minutes, and then floated on to the surface of some turpentine, where they must be allowed to remain until nearly or quite transparent, when they are to be removed to glass slides on which a little Canada balsam has been previously dropped. Examined under the microscope, the preparations will probably present, at this time, but little trace of structure; but if set aside for some time and occasionally treated with a little turpentine and Canada balsam, the cells and fibres reappear, presenting a beautiful appearance. Before the preparations are finally covered they should be examined by the microscope at intervals.

The principle of this method is this: to replace the spirit by the turpentine, and this by Canada balsam, without drying the sections.

Another plan consists in steeping the sections of the hardened structure in the carmine solution for twenty-four hours, then drying them slowly, and afterwards washing them in turpentine, and mounting in Canada balsam rendered thin by the addition of turpentine or benzole.

The illustration represents a section of the spinal cord prepared in the manner directed, by treatment with chromic acid, and subsequently by turpentine and balsam.

Fig. 134.

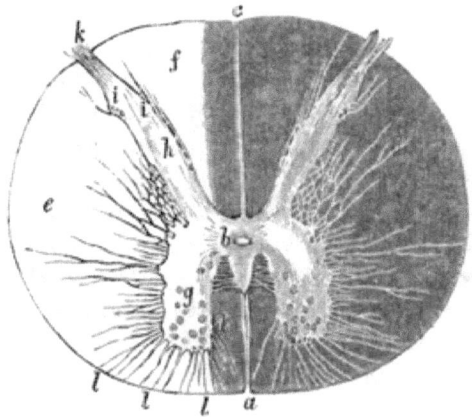

Transverse section of human spinal cord, close to the third and fourth cervical nerves; magnified ten diameters (from Stilling). *f.* Posterior column. *i, i.* Gelatinous substance of the posterior horn. *k.* Posterior root. *l.* Supposed anterior root. *a.* Anterior fissure. *c.* Posterior fissure. *b.* Grey commissure, in which a canal is contained, which, according to this writer, extends through the length of the cord. *g.* Anterior horn of grey matter containing caudate vesicles. *e.* Antero-lateral column (from *k* to *a*).

The distribution of the bloodvessels can only be studied in transparent injections, of which very beautiful examples may be obtained of the different dealers in microscopic specimens.

OF THE EYE.

As the various appendages to the visual organ—the muscles, glands, and tarsal cartilages—differ in no respect from similar tissues in other parts of the body as to the method of examination, it will not be necessary to repeat directions that have already been given. One appendage, however, being peculiar to the eyes of animals, requires especial notice. In the horse, ox, sheep, and many others, there exists at the inner canthus a semilunar expansion of cartilage, connected by means of a long stem with the fatty tissue at the bottom of the orbit, and so arranged as to be protruded over the front of the eye when that organ is drawn inwards by the action of the retractor muscle. This cartilaginous structure is termed the *cartilago nictitans*. It may easily be dissected out from the eye of an ox or sheep for the purpose of examination.

Under the microscope the ordinary groups of cartilage cells will be observed imbedded in a granular matrix.

Proceeding to the study of the eyeball and its contents, the cornea and sclerotic ought first to be examined in sections of dried preparations.

For this purpose take an eye of an ox or sheep, dissect off all the surrounding tissues, and cut away the posterior third of the globe to allow the contents to escape. The shell is then to be slowly dried until sufficiently firm to bear the pressure of the razor. Sections are to be made by first dividing the preparation across the centre of the cornea, and then cutting thin slices from the edge so as to include portions of the sclerotic and cornea.

Steeping these sections in carmine solution for a few hours not only adds to their beauty, but renders the several parts much more distinct.

The specimens may afterwards be dried and mounted in Canada balsam, or examined in a moist state on the slide in the usual way.

The drawing represents a section of the sclerotic and cornea prepared according to the method described, and shows the fibrous structure of these parts with the conjunctiva and the membrane of Demours.

Fig. 135.

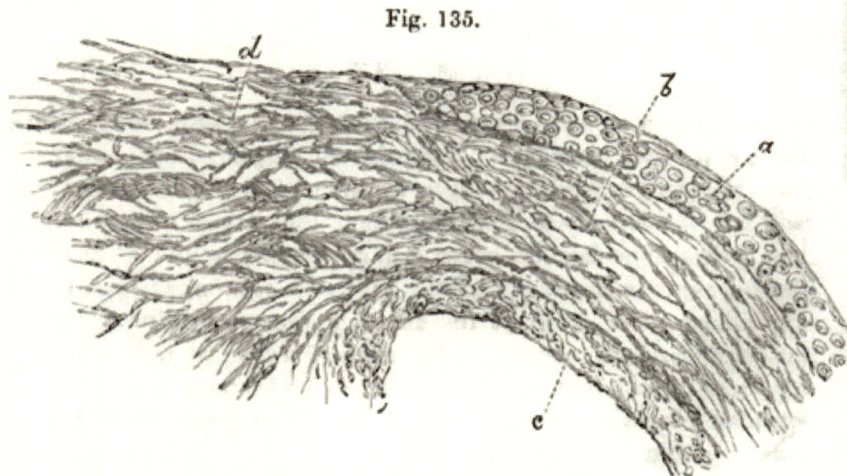

Section of cornea and sclerotic. *a.* Conjunctival epithelium. *b.* Fibrous structure of sclerotic and cornea. *c.* Membrane of Demours.—Magnified 200 diameters.

The *conjunctiva* covering the cornea is composed of layers of elongated and roundish cells.

The membrane of *Demours*, or *membrana humoris aquei*, is attached to the inner surface of the cornea and lines the anterior chamber, in which the aqueous humour is contained.

These parts are seen very distinctly in sections that have been steeped in carmine solution, and are readily distinguished from the fibrous structure of the cornea.

The epithelium on the surface of the cornea can be readily examined by scraping the surface of a perfectly fresh eye, and placing the collected matter in a drop of water on the slide, covering it in the usual way, and examining the object with a high power.

Portions of the sclerotic may be teazed out in a drop of water on the slide, and examined with a high power. The student will recognise the characteristic fibres of connective tissue. A drop of acetic acid placed at the edge of the covering glass will render apparent the few fine elastic fibres which are intermixed with the white.

A small portion of the cornea similarly treated will prove to be also fibrous in structure; the fibres, however, are extremely minute and are arranged in parallel bundles.

The vessels and nerves of the cornea are best studied in perfect specimens in the following manner. Select the eye of a young white rabbit or lamb, and cut off the cornea by a circular section with scissors close to the ciliary ring, place it on the glass slide with a little of the aqueous humour, and cut several slits round the edges of the ciliary ring, in order that the cornea may lie flat. If necessary it must be divided into three or four portions. When the structure is found to lie perfectly flat apply the covering glass, and proceed to examine first with the low power in order to discover the nerves and vessels, which may afterwards be traced towards the centre of the cornea by means of a higher objective. If the epithelium is at all cloudy a drop of caustic soda should be added to clear it. The nervous trunks are mostly dark in colour at the border of the cornea, but they become much less distinct towards the centre, and hence are difficult to trace.

The illustration will convey an idea of the appearance the observer should perceive if his specimen has been well pre-

Fig. 136.

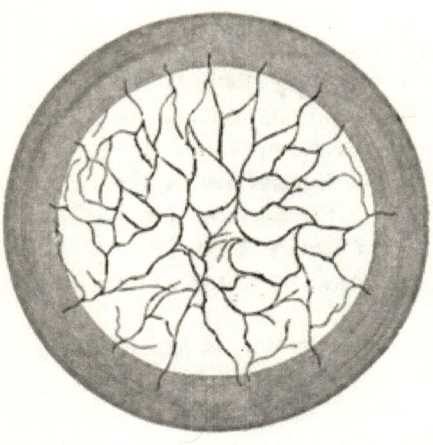

Nerves of the cornea of the rabbit in their coarser ramifications. The distance to which the dark-bordered tubes extend in the trunks is here expressed by the darker portions of the lines.

The bloodvessels of the cornea will be found only at the border, and in the adult subject they are very sparingly distributed.

The *choroid coat* is a vascular tunic extending from the entrance of the optic nerve to the iris, with which it is continuous.

Besides the bloodvessels, which are exceedingly numerous, the choroid contains numerous pigment cells of peculiar form, which may be examined by cutting off a small portion of the coat, teazing it out in a little water on the slide, and applying the covering glass in the usual way.

The specimen will present under a high power a number of the cells represented in the next illustration, besides some epithelial cells filled with pigment as well as fibrous tissue and portions of vessels.

The cells existing in the substance of the choroid are distinguished by long caudate processes; they are filled with pigment granules, and possess a nucleus which is usually light, in contrast with the dark material surrounding it.

Fig. 137.

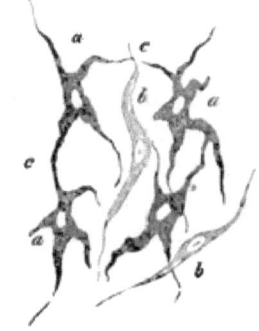

Cells from the stroma of the choroid. *a.* Pigmentated cells. *b.* Pigmentless fusiform ones. *c.* Anastomoses of the former. Of man.—Magnified 350 times.

The epithelial cells lining the choroid can be easily separated by lightly scraping the inner surface of the tunic; or, better still, touching it with the finger, and transferring the particles of dark colouring matter to a drop of water on the slide, applying the covering glass in the usual way, and viewing the specimen under a high power.

The cells are hexagonal in shape, and are filled with dark pigment granules, which are also seen scattered about the field, in size about $\frac{1}{10000}$ of an inch long. See p. 33, fig. 24.

The *tapetum lucidum*, which is peculiar to the eyes of some animals, occurs as a brilliantly metallic coloured layer situated at the upper and posterior part of the interior of the eye, extending from the entrance of the optic nerve upwards and forwards to the point of termination of the retina. A small portion of the brilliant surface should be lightly stripped off the choroid, placed in a little water on the slide, carefully teazed out, and covered with the thin glass.

Viewed with the high power, the tapetum proves to be composed of fine fibres of connective tissue arranged in bundles which interlace in all directions. Some pigment cells and stroma cells of the choroid will often be present, but they are to be considered as accidental.

The surface of the tapetum is covered with a layer of extremely delicate epithelium; the cells of which resemble in shape the pigment cells of the choroid, but are devoid of colour, in fact, are so pale that they are not visible in a strong light.

In order to examine them it is necessary to take the lightest scraping from the surface of the tapetum, place the collected matter in a drop of water on the slide, cover in the usual way, and examine the preparation with the high power.

The best way of illuminating this specimen is to turn the mirror a little to one side in order that the light may be somewhat obliquely thrown upon the object. The cells will appear with very delicate outlines, distinct rather large nuclei, and faintly granular contents.

So much of the success of this observation depends upon the method of illumination that the student must not be discouraged if he should fail at first to see the object at all.

The bloodvessels of the choroid and iris can be most satisfactorily studied in injected preparations. Their arrangement is beautifully shown in the drawing given below.

Fig. 138.

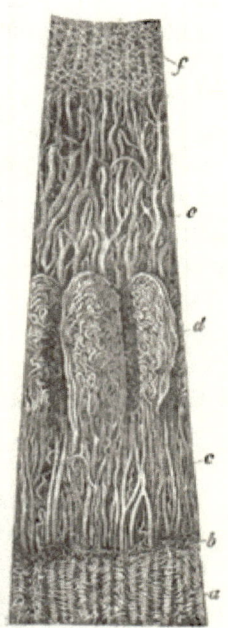

Vessels of the choroid and iris of a child, after Arnold, seen from within, and magnified 10 times. *a.* Capillary network of the posterior section of the choroid, terminating at the *ora serrata*, *b*. *c.* Arteries of the *corona ciliaris*, supplying the ciliary processes *d*, and partly proceeding to *e*, the iris. *f.* Capillary network of the inner surface of the pupillary border of the iris.

The nerves of the iris and many of the vessels may be seen by removing the part from a perfectly fresh eye of a white rabbit, placing a portion upon the glass slide, adding a little dilute caustic soda, and using a low power.

Fig. 139.

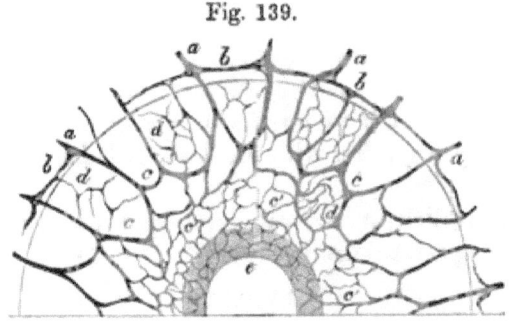

Nerves of one half of the iris of a white rabbit; magnified 50 times. *a.* Nervuli ciliares. *b.* Anastomoses of the same at the border of the iris. *c.* Larger arcuate connections of the same in the iris. *c'.* Finer networks of the same in the inner parts. *d.* Terminations of separate nervous filaments. *e.* Sphincter pupillæ.

On the edges of the pupillary opening in the eyes of the horse, ox, sheep, and many other animals, are peculiar little black bodies (*corpora nigra*) which can be seen in the living animal and are not uncommonly mistaken for products of disease. They are fewer in number and larger on the superior edge than on the inferior.

These little bodies may be examined in the eyes of the sheep or ox, obtained from the butcher. The cornea having been removed, one of the corpora nigra is to be taken in the forceps, cut off with scissors and transferred to the slide for the purpose of being teazed out in a little water, and covered in the usual way.

Under the high power the structure is found to be entirely composed of pigment cells, exactly the same as those from the inner surface of the choroid.

The retina, which is a delicate transparent membrane, extends from the entrance of the optic nerve to the outer or posterior edge of the ciliary body, where it properly terminates in an undulated or jagged edge, the *ora serrata*. From this point there is continued a layer of nucleated cells reaching to the tips of the ciliary processes, constituting the *pars ciliaris retinæ*. A careful examination of the retina will lead to the observation of several coats or layers of very peculiar structure.

158   HISTOLOGY.

In order to see these tissues it is necessary to make vertical sections, which may be done in the following way. Remove a fresh retina carefully and spread it out on a piece of glass, and add a strong solution of chromic acid. After a few minutes sections can be made with a razor or sharp scalpel in any required direction. The best method is to cut rather obliquely, beginning with a moderate thickness, and tapering the section off to a very thin edge. In this way one part is almost certain to be sufficiently fine. The specimen is to be placed on the slide, and examined in the ordinary way, first with the low, and afterwards with the high power.

In the illustration the several parts which the observer should see in a good specimen are indicated.

Fig. 140.

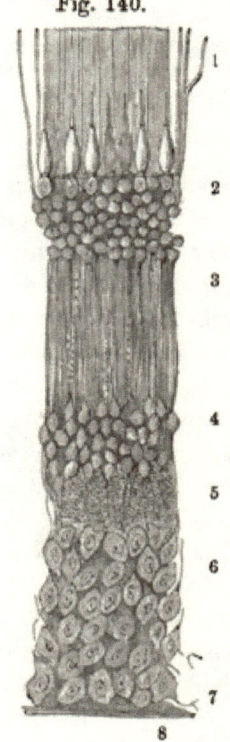

Perpendicular section of the human retina, made six lines anterior to the entrance of the optic nerve; magnified 350 times. 1. Bacillar layer. 2. Outer granular layer. 3. Intermediate granular layer. 4. Inner granular layer. 5. Finely granular grey layer. 6. Layer of nerve-cells. 7. Fibres of the optic nerve. 8. Membrana limitans.

An examination of the separate elements can only be made by teazing out portions of the fresh retina in a little aqueous humour or dilute solution of glycerine, on the glass slide, covering the preparation, and viewing it with the high power.

The elements of the bacillar layer or *membrana Jacobi* will appear as peculiar rods, as shown in the illustration.

Fig. 141.

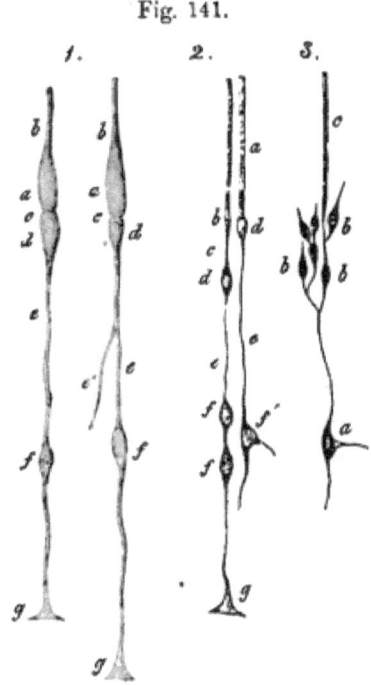

Elements of the bacillar layer in connection with the fibres of *Müller*. From the human retina, magnified 350 times. 1. Cones, with fibres of *Müller*. *a.* Thicker part of the cone, or proper cone. *b.* Rods upon the same, one longer than the other. *c.* Circular line at the inner end of the cone. *d.* Nucleated swelling (cell-body) of the same already in the outer granular layer. *e.* Fibre of *Müller*, into which it is continued. *e'.* Lateral process from one of these fibres, passing inwards. *f.* Granule (cell) of the inner granular layer. *g.* Internal termination of the fibre of *Müller*. 2. Rods with Müllerian fibres. *a.* Rod. *b.* Transverse line at its inner extremity. *c.* Commencement of the filament of *Müller*. *d.* Granules of the outer granular layer, one seated close to the rod. *e.* Fibres of *Müller* in the intermediate granular layer. *f.* Inner granules. *f'.* One of them with a lateral process. *g.* Internal extremities of the fibres of *Müller*. 3. An inner granule. *a.* With three processes, of which the external one gives off branches, and supports several other granules, *b*, together with rods, of which only one is figured.

160 HISTOLOGY.

The nerve cells sometimes present numerous processes, as shown below.

Fig. 142.

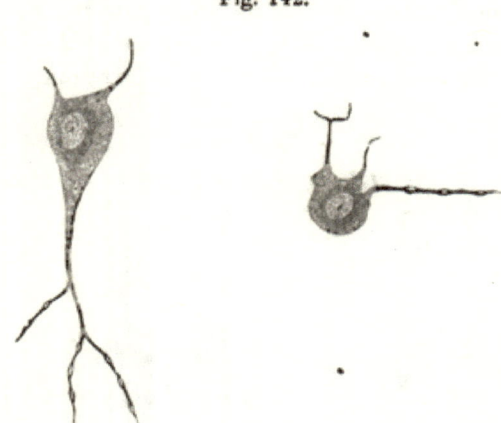

Two nerve cells from the human retina, magnified 350 times. The smaller with two processes outwards, and only one varicose nerve fibre arising from it; the other with a dividing process which passes into three nerve fibres, and two similar processes torn off.

The Crystalline Lens.—In perfectly fresh specimens the delicate capsule may be readily separated from the lens, and its epithelial lining demonstrated; but the structure of the lens itself is best studied in specimens that have been boiled, or otherwise hardened by immersion in alcohol or chromic acid. A very good plan is to allow a fresh lens to remain on a slip of glass exposed to the air for a day or two, until it becomes sufficiently firm to be cut with a razor. In this way the transparency of the structure is in a great degree preserved.

From the hardened lens the concentric laminæ may be stripped off layer by layer, until we arrive at the central point.

In order to examine the structure of the laminæ, a small portion of one of them should be placed in a drop of water on the slide, and gently separated into fibres in a longitudinal direction by means of needles, then covered and examined with a high power.

A number of flattened bands will be seen; some of them quite separate, and others united together by their edges, which are denticulated as shown in the next wood-cut.

Transverse sections, which are easily made with a razor from a lens previously hardened, prove the bands or fibres to be hexagonal in form, as indicated in the drawing.

Fig. 143

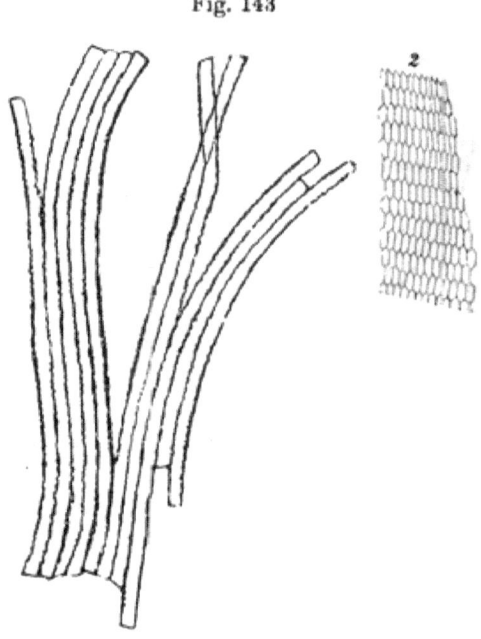

Lenticular tubes or fibres. 1. Of the ox, with slightly dentated margins. 2. Transverse section of the lenticular fibres of man.—Magnified 350 times.

The crystalline lens has been, not inaptly, compared to an onion, consisting of a series of laminæ or layers, which, in the hardened structure, may readily be stripped off, a layer at a time.

The bands of which the laminæ are composed are not of the same breadth throughout, but taper towards the extremities in the same way as the spaces between the meridian lines of a geographical globe, which are broad at the equator and gradually become narrower as they converge to the two opposite poles.

This arrangement is seen in the most simple form in the lenses of birds and some fishes as depicted in the next drawing.

Fig. 144.

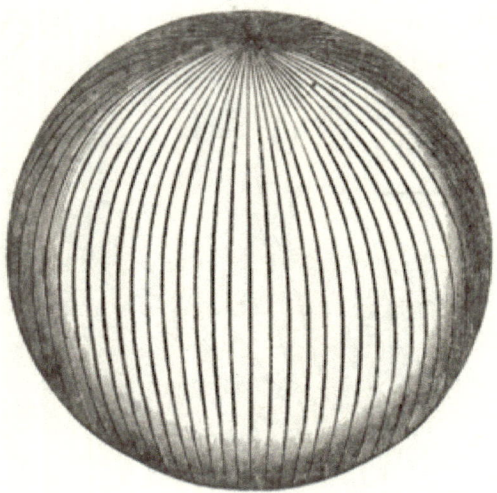

SIMPLEST FORM OF CRYSTALLINE LENS.

A more complicated plan is found among some mammalia, as the rabbit, hare, &c., as well as in certain fishes and reptiles, the centres or poles on each side being crossed by a transverse line or septum, as in the next figure.

Fig. 145.

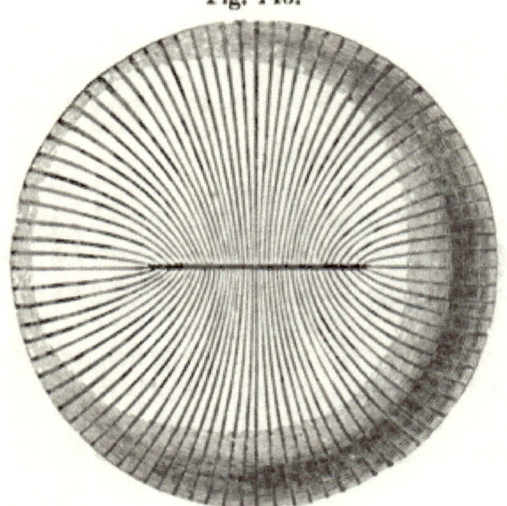

LENS WITH TRANSVERSE SEPTUM.

Among mammalia in general there are three lines or septa diverging from the poles on each side; these may be distinguished very easily in a lens that is just beginning to lose its transparency, and appear exactly as represented beneath.

Fig. 146.

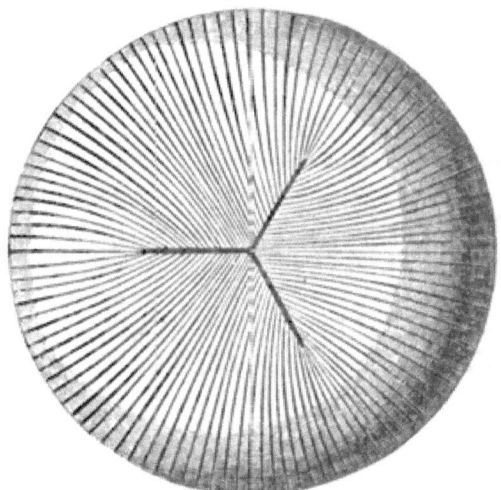

LENS WITH THREE SEPTA.

The central portion of the lens, that part which remains after all the laminæ are stripped off, is depicted in the next drawing, from the eye of the *Salamander maculata*.

Fig. 147.

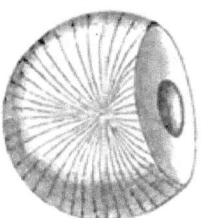

Central point of the crystalline lens of the *Salamander maculata*. A small portion is cut away, showing the centre to be hollow.

The vitreous body, which occupies the space between the lens and the retina, is enveloped in a fine membrane, the hyaloid, which, according to Kölliker, proceeds to the border of the lens, in order to coalesce with it. During this course it divides into two lamellæ; the anterior division extends forward to become attached to the lens a little in front of its border, and forms the zonule of Zinn.

The posterior lamella unites with the capsule of the lens, behind its border. The space between the two lamellæ is the canal of Petit.

The vitreous body is perfectly transparent and colourless, and contains no microscopic elements of importance.

THE INTERNAL EAR.

The parts composing the external ear, consisting, as they do, of integument, cartilage, muscle, bone and mucous membrane, present no special points for microscopical study, nor will the method of examination differ from that previously recommended for the study of the same structures in other parts of the body.

The tissues contained in the labyrinth of the ear, however, are exceedingly delicate and beautiful, but especially difficult to prepare.

Fig. 148.

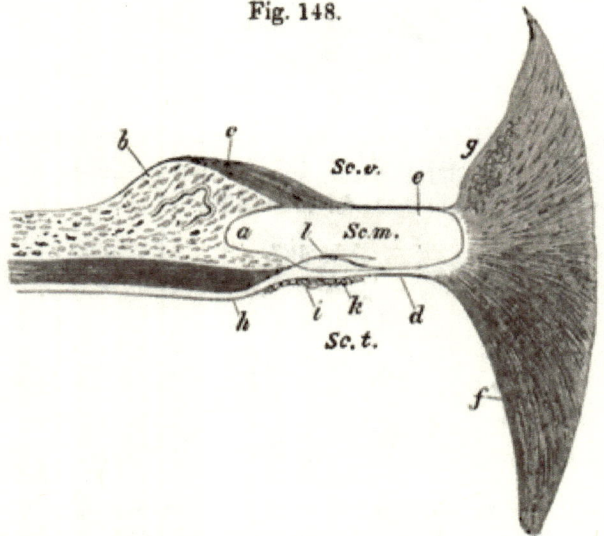

Transverse section through the spiral lamella of the first turn of the cochlea. From the ox; magnified 100 diameters. *Sc. t. Scala tympani. Sc. v. Scala vestibuli. Sc. m. Scala media. a. Sulcus spiralis. b.* Teeth of the first series. *c. Membrana Corti,* its thicker part. *d. Membrana basilaris. e.* Corti's membrane, its thinner part. *f. Lig. spirale. g. Stria vascularis. h.* Nervous expansion in the zona ossea. *i. Vas spirale internum. k.* Layer of corpuscles of connective tissue, with varicose processes from them. *l.* Organ of Corti, only just indicated.

Specimens from animals recently killed are necessary, and instead of water, serum or a solution of sugar must be employed as the fluid medium on the glass slide.

Sections of the lamina spiralis may be made from preparations hardened in chromic acid and afterwards treated with hydrochloric acid, or the entire cochlea may be softened in moderately strong hydrochloric acid, and afterwards washed and laid in a bed of gelatine, when it may be cut with a razor.

In this method of preparation it is necessary to liquefy the gelatine by the agency of heat, and then to introduce the structure of which sections are required, and when the mass becomes cold it will bear the pressure of sharp scissors or a razor.

OF THE NOSE.

THE structures which unite to form the olfactory apparatus are bone, cartilage, muscle, and mucous membrane; and specimens of these parts are to be prepared for examination according to the methods previously described. It will be observed that the osseous structure of the ethmoid, at least at its thinnest part, is devoid of Haversian canals, consisting only of a matrix in which a number of lacunæ are distributed.

The cartilage of the nose possesses a finely granular matrix, and the contents of the cells are generally transparent.

Those parts of the organ of smelling which particularly interest the microscopist are the Schneiderian membrane and the olfactory nerves.

1. *The epithelial cells* covering the mucous tissue are to be examined, by lightly scraping the surface of the membrane with a scalpel, and treating the matter obtained in the usual way.

2. *Sections of the mucous membrane* are to be made with fine curved scissors from the edges of a detached portion of the fresh tissue, or with a razor from a dried preparation.

The section should be placed on the slide with a little serum or white of egg, covered, and examined with a low power.

In a good specimen the structures shown in the drawing will be seen.

Fig. 149.

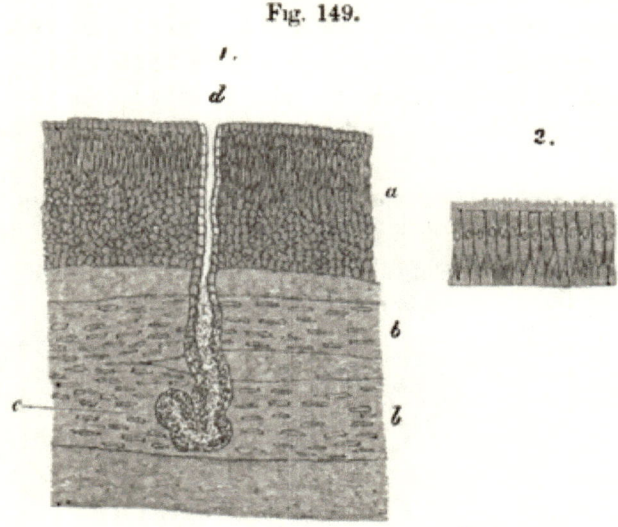

From the nasal mucous membrane of the sheep; magnified 150 times. 1. Transverse section of the mucous membrane, from the proper olfactory region. *a.* Epithelium without cilia. *b.* Olfactory nerves, with a dividing pale nucleated bundle. *c.* Gland of *Bowman*. *d.* Aperture of the same. 2. Ciliated epithelium of the Schneiderian membrane.

The olfactory nerves may be studied in the nose of the ox or sheep. The cavity of the nostrils being laid open, a small piece of nerve should be dissected up from the olfactory region, and placed, if possible, in a little vitreous humour, in preference to water, on the slide, and well teazed out, the covering glass being carefully applied in the usual way, and the object viewed with a high power. Fresh specimens should be selected for examination, as these tissues rapidly undergo change.

As will be seen in the illustration, the olfactory nerves of mammalia do not contain any white fibres, but consist solely of flat tubes with granular contents.

The tubes are united together by the intervention of white fibrous tissue. In the engraving, which is taken from a portion of the olfactory nerves of the ox, the connective tissue is not represented.

THE NOSE.

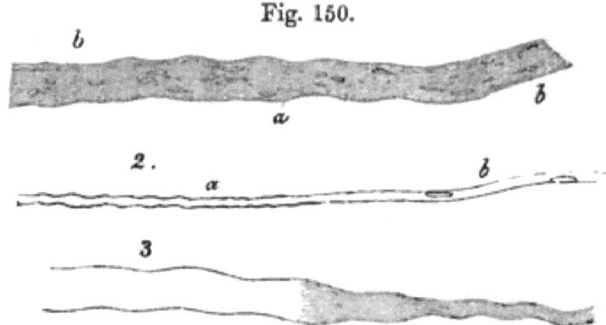

Fig. 150.

Olfactory tubes of the ox; magnified 350 times. 1. A thick grey tube. Envelope of the same. *b.* Effused contents with nuclei. 2. A fine dark-bordered tube, *a*, continuing from one of the foramina cribrosa into a pale nucleated fibre, *b.* 3. The empty envelope of a grey tube, at one of its extremities appearing collapsed and like a fibre.

Mucus which is secreted by the nasal membrane has the least admixture of foreign particles, and is therefore to be preferred for microscopic examination. A small quantity is to be removed from the surface of the membrane by means of a camel-hair pencil, transferred to the slide without the addition of water, then covered in the usual way, and viewed under a high power.

The mucous secretion consists of a transparent viscid fluid, containing a variable number of corpuscles, which present different characters, not only in the mucus from the membranes of various parts, but even in the same specimen. Some are larger than the colourless corpuscles of the blood, circular or oval in shape, with defined outlines, and possess two or three nuclei, which are faintly seen through the cell-wall. Others are much smaller, irregular in outline, and present a granular surface (see fig. 163 B, part II.).

Some epithelial cells will always be observed, and often in considerable numbers.

A drop of acetic acid applied to the edge of the covering glass causes an immediate coagulation of the fluid in long fibres which entangle the corpuscles. The nuclei are at the same time rendered very distinct.

ORGANS OF GENERATION.

The parts most interesting to the microscopic observer are, in the male, the testicles with their ducts, and the spermatozoa in the various stages of development; and in the female, the ovaries with the Graafian follicles, and the corpus luteum.

The other portions of the generative system, including the mucous membranes and muscular structures of the uterus, as well as the bloodvessels and nerves, are to be studied according to the methods previously recommended for the preparation of similar tissues in other parts.

In order to examine the seminal tubules, a testicle must be cut across and a scraping taken from the cut surface, placed in a drop of water on the slide, and teazed out, the covering glass being applied and the object viewed first with the low and afterwards with the high power.

The preparation, if successful, should present the characters depicted in the drawing.

Fig. 151.

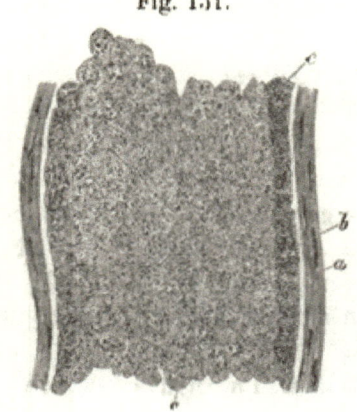

Seminal tubule of man; magnified 350 times. *a.* Fibrous coat with longitudinal nuclei. *b.* Clear border, indicating, probably, a basement membrane. *c.* Epithelium.

A seminal tubule will be found to consist of a thick fibrous coat lined with epithelium, placed on a basement membrane.

Contained in the tubules are numerous cells, which, in young animals, are small and clear, hardly to be distinguished from epithelium. In sexually mature subjects the cells are filled with nuclei, and are looked upon as the precursors of the seminal fluid. In the course of examination of the contents of the tubules the sperm cells will be seen with the spermatozoa in various stages of development, as shown in the illustration.

Fig. 152.

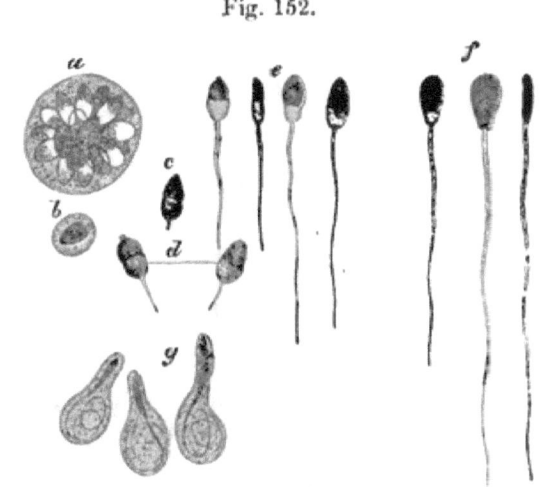

Development of the spermatic filaments of the bull. *a.* Sperm cyst, and *b*, sperm cells, containing nuclei, which exhibit a darker anterior and a clearer posterior part. *c, d.* The nuclei developing into spermatozoa. *e.* Further stages. *f.* A filament almost complete, but still showing a little of the posterior clear zone in its body; also two perfect spermatozoa from the epididymis, one seen from the surface, the other edgewise. *g.* Sperm-cells becoming pyriform, and about to liberate the contained spermatozoa.—Magnified 450 diameters.

To observe the spermatozoa in motion, a little of the fluid must be pressed from the vas deferens, or seminal vesicles, of an animal recently killed—a frog answers admirably for this purpose. Their movements may also be seen in the milt of a male fish at spawning time. A little of the milt may be gently pressed from the live fish without injuring it, and received in a drop of water on the glass slide; the covering glass should be immediately applied, and the object viewed with a high power.

The spermatozoa will be seen as small oval bodies moving rapidly about the field of the microscope. In specimens that were examined from the salmon and trout no filaments could be detected.

170   HISTOLOGY.

The characters presented by the human spermatozoa are well seen in the drawing.

Fig. 153.

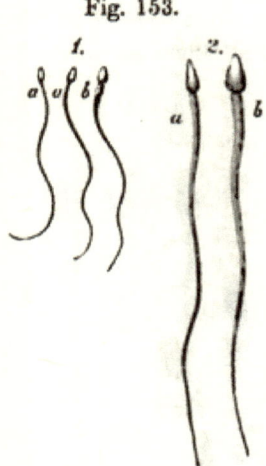

Human spermatozoa. 1. Magnified 350 times. 2. Magnified 800 times.
a. Seen from the side. *b.* From the surface

Spermatozoa taken from the various classes of the animal kingdom show considerable diversity of form. A few of the most interesting specimens are represented in the succeeding illustration.

First, those from some of the insect tribe, comprising filaments without heads.

Fig. 154.

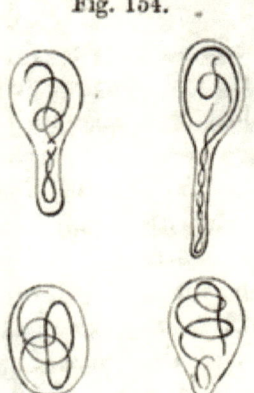

Spermatozoa in the interior of the vesicles of development of *Nepa cinerea.*

Second, spermatozoa with a rounded head and a filamentous tail extending from it, as in the common perch and some other fishes.

Fig. 155.

Spermatozoa of *Perca fluviatilis*.

Next, those with oblong heads, found in the domestic cock.

Fig. 156.

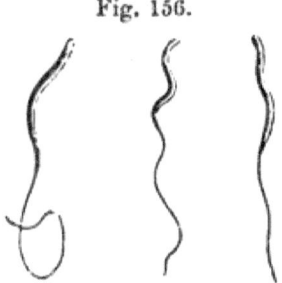

Spermatozoa of the cock (*Gallus domesticus*).

In some birds, as sparrows and finches, the heads of the spermatozoa are wavy, as seen in the drawing.

Fig. 157.

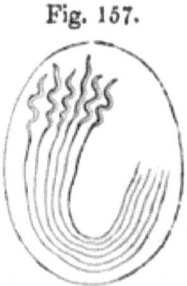

Mother cell with a bundle of spermatozoa from *Fringilla domestica*.

In rats and mice the spermatozoa have very long tails and heads of very peculiar form.

Fig 158.

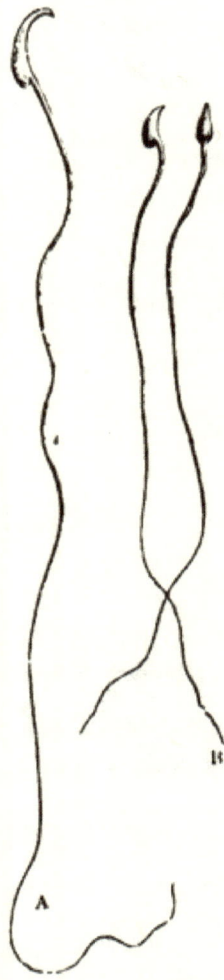

A. Spermatozoa of the rat.   B. Of the common mouse.

The Ovaries of the female are composed of a dense fibrous stroma, in which numerous small pellucid bodies (Graafian follicles) are imbedded. The entire organ is invested with a fibrous coat and a covering of peritoneum.

On making a section of an ovary, the Graafian vesicles will appear most numerous towards the circumference.

THE OVUM.

A single vesicle may be examined by dissecting a specimen from the ovarium of a cow, placing it in a little water on a slide, and applying sufficient pressure with the covering glass to flatten it slightly. The structure of the capsule and the portion of the ovum within it may sometimes be made out in the fresh specimen, but sections of an injected and hardened ovarium are much to be preferred.

To obtain a single ovum for examination, a large Graafian follicle should be punctured and the fluid allowed to flow on to a piece of glass. The ovum should then be sought with a lens or under the low power, and, when found, covered with the thin glass and examined with the high power.

The drawing indicates what the student should see in a good specimen.

Fig. 159.

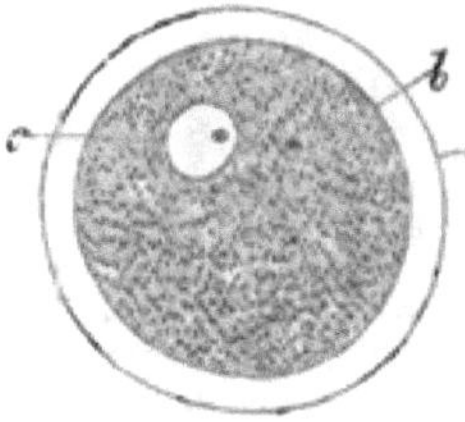

Human ovulum, from a middle-sized follicle; magnified 250 times. *a*. Vitelline membrane (zona pellucida). *b*. Outer limit of the yelk, and, at the same time, boundary of the vitelline membrane. *c*. Germinal vesicle, with the germinal spots.

After the escape of the Graafian follicle from the ovarium, the space it occupied is filled up with a yellowish-white substance, termed the corpus luteum. The masses thus formed are very apparent upon the cut surface of the ovarium, particularly from an animal that has been frequently pregnant.

The structure of the true corpus luteum, which in every respect except size resembles the false, may be demonstrated by teazing out a small portion in water upon the slide, and covering it, previous to examination with a high power. The elements are found to be nucleated cells, which become partly transformed into imperfect connective tissue.

Sections of these bodies may be made in fresh specimens with the double knife, or from those hardened in chromic acid with a razor. The section should be placed in water on the slide and covered in the usual way, and examined with the low power.

The various parts which should be seen in a good section are presented in the illustration.

Fig. 160.

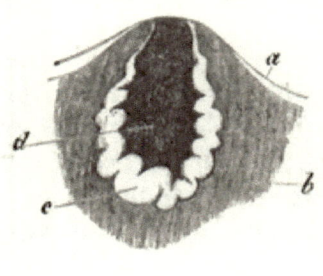

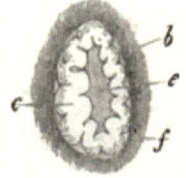

Sections of two *corpora lutea*, of natural size. The upper one represents a specimen eight days after conception. The lower object is from an ovary during the fifth month of pregnancy. *a.* Tunica albuginea of the ovary. *b.* Stroma of the ovary. *c.* Thickened and plaited fibrous coat of the follicle (inner layer). *d.* Blood coagulum, and *e*, decolorised blood coagulum within the preceding. *f.* Fibrous envelope which limits the corpus luteum.

THE MAMMARY GLANDS, or lacteal glands, exist in the male animal as rudimentary organs, but in the female they are developed fully and possess the function of secretion.

In structure the lacteal glands correspond to the pancreas and salivary glands, being composed of lobes formed by the union of small lobules, which are constituted of gland vesicles. These elements are united by white fibrous tissue.

The structure of the mammary glands is best studied by teazing out portions on the slide, and in sections of hardened and injected preparations. Specimens which are prepared by being hardened in alcohol or by boiling may be rendered sufficiently transparent by immersion in glycerine.

The arrangement of the lobules is seen in the illustration.

Fig. 161.

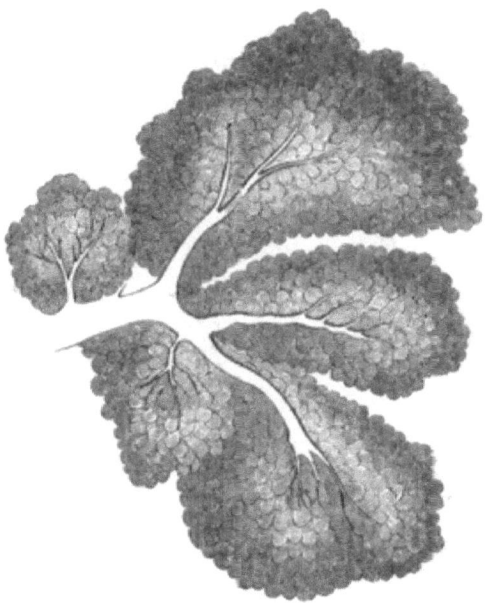

Some of the smallest lobules of the lacteal gland of a puerperal woman, with their ducts. After Langer.—Magnified 70 times.

Milk is to be examined in the manner adopted for other fluids, a drop being placed in the centre of the slide and covered with the thin glass. Care must be taken that the film is not too thick, otherwise the elements will not be distinct. When properly prepared the fluid will be found to consist of a transparent medium, in which numerous oil globules are floating. In the milk secreted immediately before and after parturition, a number of compound corpuscles, called colostrum corpuscles, are found. Some little care is necessary in the examination of milk, to avoid confounding small aggregations of milk globules with colostrum corpuscles, which are surrounded by a fine membrane, or cell wall, as shown in the illustration, and particularly at c. The aggregations of the ordinary milk globules are usually larger than the colostrum corpuscles, and their outline is less regular; but the absence of any defined wall round the mass will be the most reliable and characteristic point of difference, although it is most probable that these aggregations of milk globules and the colostrum corpuscles have a common origin, as the secretion of milk commences by the formation of a fluid, in the extremities of the gland follicles, containing numerous cells filled with fat globules, which are set free by the bursting of the cell wall, even before the secretion leaves the lacteal ducts.

The illustration includes this occasional element as well as the ordinary milk globules.

Fig. 162.

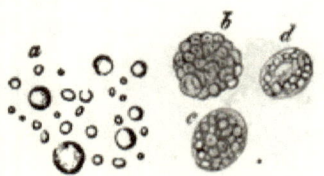

Morphological elements of the milk; magnified 350 times. *a*. Milk globules. *b*. Colostrum corpuscles. *c, d*. Cells with fat globules, from the colostrum, the one, *d*, with a nucleus.

MORBID HISTOLOGY.

Microscopic investigation of the animal tissues would hardly repay the student for the expenditure of so much valuable time, if it did not lead to far more important results than the acquirement of a knowledge of the nature and arrangement of the elementary constituents of the body.

In reality the microscope has yet to be universally recognised in its true position as an indispensable aid to diagnosis, extending but not supplanting the ordinary means of observation.

Valuable as the instrument is to the histologist in the dissecting-room, it is of infinitely greater importance to the physician and surgeon seeking to discover the causes and remedy the effects of disease.

Morbid secretions when examined furnish evidence of the utmost value, only accessible to the microscopist.

Internal growths may be often reached by the exploring needle, while external tumours present no difficulties in the way of an examination, by which their nature may be determined before an operation has rendered an error irremediable.

To initiate the student into the method employed in the examination of morbid products, and to furnish him with some general facts, an acquaintance with which will much facilitate his investigations, are the special objects of the second part of the course.

Those diseases only will be considered that are of most frequent occurrence; and, as far as possible, the illustrations will include only typical forms, avoiding any reference to those minute gradations which, gliding imperceptibly one into the other, tend rather to perplex than to instruct.

As far as explanations of the various morbid states are concerned, they must of necessity be brief, hence statements will often assume rather the dogmatic than the argumentative form, as better calculated to impress upon the mind those great general principles, which are indispensable to the successful study of this branch of the subject.

Morbid action manifests itself under three well-marked conditions:—

1. In the abnormal growth of an already existing normal tissue—Hypertrophy.

2. In an abnormal decrease of an existing normal tissue—Atrophy.

3. In the modification of the elements of an existing normal tissue—Degeneration.

4. In disease neither new histological elements, nor new functions are created; but existing tissues are modified or misplaced, and existing functions become disordered.

Hence the presence of a normal structure in an abnormal situation will constitute disease. A piece of bone occurring in a gland would be designated a morbid growth (osteoid); a portion of glandular tissue growing in the medullary canal of a bone would be considered an abnormal product (adenoid). Nevertheless, in both cases, the piece of bone on the one hand, and the portion of gland on the other, might possess all the characters of healthy tissue. It is not only, therefore, the nature, but also the position of a deposit, that determines the title of morbid.

5. All tumours grow in the direction of the least resistance, consequently their forms depend upon the extent and position of the pressure to which they have been subjected. Where there is entire freedom from pressure on all sides, the form of the tumour is globular, as is the case when they grow into cavities.

6. All kinds of tumours are liable to occur in any situation, external or internal. Osseous growths may be developed in muscle, epithelial in nerve, cystic in bone, &c.

7. All tumours are liable to inflame, and take on rapid growth.

8. All tumours may be multiple, as instanced in a general tendency to ossific deposits: multiple sebaceous tumours on the head; fibrous tumours in the uterus, in enchondroma; fatty tumours; and cancerous deposits in the liver, spleen, and lung.

9. All growths are liable to recur.

10. The more fibrous the less likely to extend or recur; the more cellular, the more likely to extend or recur.

11. All tumours are liable to alteration in course of their growth. A fibrous tumour may be softened or become cystic, or undergo the process of hardening or ossification. A cancer may undergo fatty degeneration (it will never become converted into a fatty tumour).

12. Benign tumours show little tendency to infiltrate neighbouring tissues and affect glands.

13. Malignant tumours alone affect glands, and cause 'cancerous cachexia.'

14. The more morbid growths differ from the tissues in which they occur, the more dangerous their character.

With these axioms impressed upon his memory, the student may proceed to investigate the structure of morbid growths and deposits, still adhering to the methods with which he is already familiar.

One rule, however, must take precedence of all others in the microscopic investigation of morbid deposits, solid or fluid, viz.: to examine the specimen as soon after death as possible. Under the influence of disease both solids and fluids are prone to change, and it often happens that a morbid growth, rich in characteristic pathological elements, becomes reduced in a day or two to a mere granular mass, whose nature it is impossible to define.

Some structures, as fibrous tumours, may remain unchanged for forty-eight hours, and some fluids, urine for example, must often be allowed to rest, in order that any adventitious matters may be deposited; but under no circumstances should any unnecessary delay be allowed to take place, especially in warm weather, and more particularly in the case of malignant growths, which undergo changes so rapidly that an examination, even after the lapse of twenty-four hours, will often be very unsatisfactory; at any rate, an opinion based upon such an observation should be accepted with extreme caution.

MICROSCOPIC EXAMINATION OF MORBID FLUIDS.

Pus.—Laudable pus is a thick creamy fluid, of an opaque yellow or slightly greenish colour, having an alkaline reaction when perfectly fresh, but becoming acid after a short time.

When allowed to remain at rest in a vessel it becomes separated into a serous fluid and a sediment.

Under the microscope pus is seen to contain a number of corpuscles with a granular surface, having two or three nuclei, generally visible through the cell wall. Besides the pus corpuscles, there are commonly large granular corpuscles, and also granular matter, with fat granules.

A drop of acetic acid or caustic potash placed at the edge of the covering glass will cause the walls of the cells to become transparent and render the nuclei very distinct.

Under certain conditions, however, the appearance of the corpuscles is considerably modified. Sometimes their outlines are irregular. Often the cell walls are so opaque that the nucleus is invisible, and in many instances the corpuscles are remarkably shrunken. Occasionally blood discs, hairs, epithelial scales, and mucus cells are present.

In spite of the microscopic differences in the appearance of pus corpuscles, and notwithstanding the admixture of adventitious products, it may always be taken for granted that pus is present when the field of the microscope is seen to be crowded with granular corpuscles, nearly all of which are of uniform size and appearance, and possess double or treble nuclei, which are rendered distinct by acetic acid or caustic potash.

In the absence of the microscope pus may be confounded with serous exudation containing a quantity of dissolved fibrin, or with any fluid possessing the same general characters. With the aid of the instrument, however, such errors are impossible.

Upon the much-vexed question of the distinction between pus corpuscles, the mucus corpuscles, and the white corpuscles of the blood, it is unnecessary to enter, as it may reasonably be doubted whether any appreciable differences exist. The fluids, however, of which the cells form a part, possess certain general and microscopic characters, which, taken in their entirety, suffice to separate them from each other. But to attempt to decide from the examination of a few isolated corpuscles, whether they belong to pus, blood, or mucus, is about as rational as to seek to determine, by the inspection of a few single bricks, whether they formed part of a cottage or a castle.

The illustration represents the pus and mucus corpuscles in their natural state, and after treatment with acetic acid.

Fig. 163.

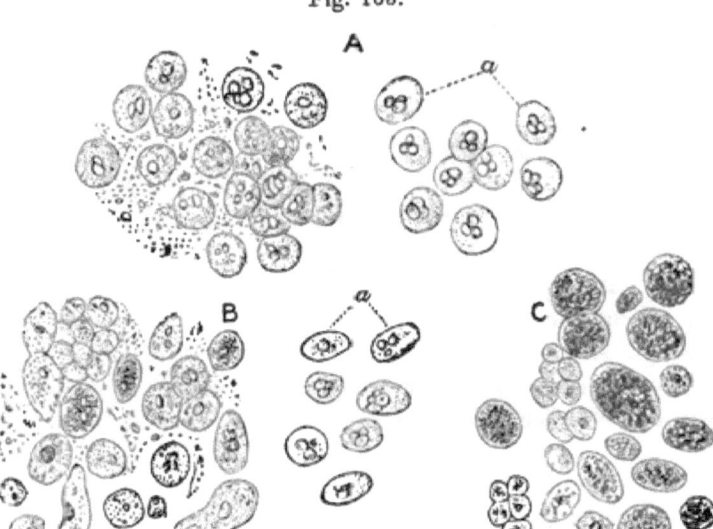

A. Pus corpuscles from an abscess. *a.* The same after treatment with acetic acid. B. Mucus corpuscles from the Schneiderian membrane. *c.* After a drop of acetic acid has been added. C. Mucus corpuscles speckled with pigment granules, from a case of chronic irritation of the lining membrane of the larynx, with expectoration of grey sputa.

Serous Fluid effused into large cavities, as the thorax or abdomen, frequently contains pus cells, and also granular corpuscles, in addition to masses of fibrous material.

Sometimes the fluid from serous cysts will be found, under the microscope, to contain scolices of some hydatid. An echinococcus cyst, for example, occurring in the region of the neck, will be very probably mistaken for a serous abscess until the fluid is evacuated and submitted to minute investigation, when the presence of scolices or fragments of them, or even of some of the hooks, will at once prove the nature of the tumour.

The secretions, and even the blood from whence they are derived, undergo certain changes, the result of disease, and frequently contain abnormal products.

Mucus from the stomach and intestines may contain vegetable organisms, both fungi and algæ. The secretion from the bronchial membrane very frequently includes the ova of entozoa, or even the parasites themselves; pus corpuscles and blood discs are also commonly seen.

182   MORBID HISTOLOGY.

Milk.—During certain diseased states, the mammary gland will contain blood discs, pus corpuscles, and granular corpuscles, which must not be confounded with colostrum corpuscles, normally present in the fluid immediately after parturition, as represented in fig. 162, p. 176.

Urine, more than any other secretion, requires the aid of the microscope for the detection of the numerous abnormal products which it frequently contains.

In the limited space assigned to this portion of the course it would be impossible to include the necessary directions for the microscopic and chemical examination of the secretion, but a few of the more common forms of urinary deposits are depicted for the guidance of the student in the course of his investigations.

Urinary Deposits.

One of the most frequent deposits in urine is the uric acid, the crystals of which occur most commonly as shown in the next illustration. For the detection of this deposit the low power is generally sufficient, as the crystals are large, sometimes perceptible to the naked eye.

Fig. 164.

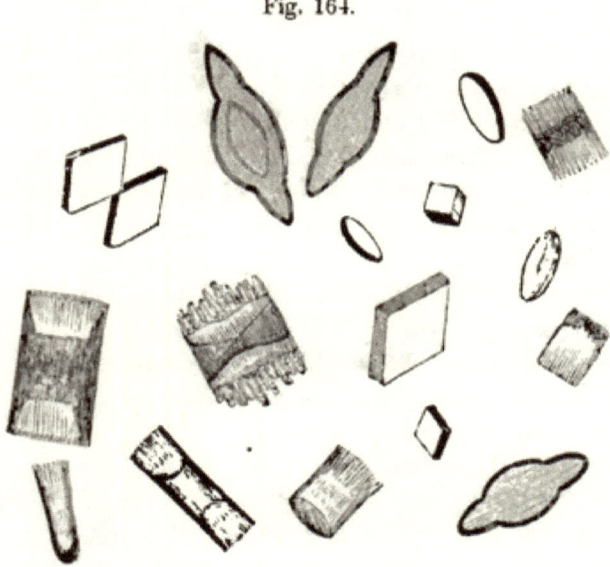

CRYSTALS OF URIC ACID.

Another not uncommon deposit is oxalate of lime, which occurs in two forms; the octahedral crystal, which is most characteristic, and the dumb-bell crystal, which is also one of the forms in which carbonate of lime occurs in the urine of the horse. The illustration shows both these forms of crystal from human urine.

Fig. 165.

Oxalate of lime from urine.—Magnified 200 times.

The triple phosphates are at once to be recognised by the presence, under the microscope, of large prisms like those shown in the next drawing.

Fig. 166.

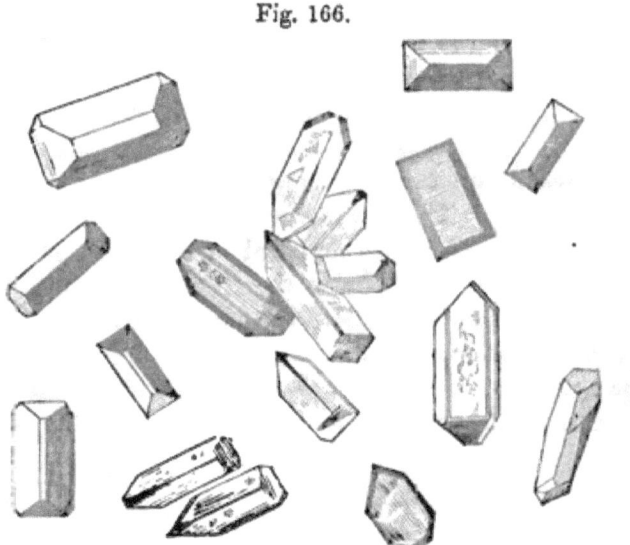

Crystals of triple phosphate.—Magnified 200 diameters.

184 MORBID HISTOLOGY.

Besides crystalline deposits there are organic products, which are frequently found in the urine, for instance, fat, fibrinous casts of the uriniferous tubes, blood, pus, and, occasionally, the ova of entozoa.

The accompanying drawing represents a number of tube casts with cylindrical epithelium and a few crystals of oxalate of lime.

Fig. 167.

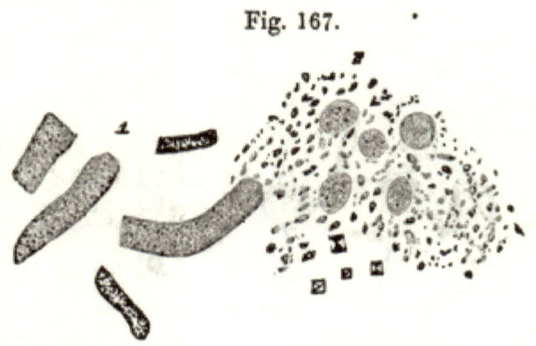

Urinary deposits from a case of organic disease of the kidneys. 1. Tube casts. 2. Cylindrical epithelium. A few crystals of oxalate of lime, with a quantity of amorphous matter, are also depicted.

Blood.—Besides the crystalline bodies described at p. 48, blood may contain other microscopic elements, whose presence in abundance may be important, for example, an excess of white corpuscles, which would be readily detected. Ova of entozoa may be discovered, and at once suggest the presence of parasites in some parts of the organism from whence the fluid has been procured.

Sometimes minute organisms are observed moving with extraordinary rapidity. These are termed bacteria and vibrines.

Bacteria are found in splenic blood of animals that have died from splenic apoplexy. Under the high power they appear as minute jointed filaments possessing great facility of motion.

Vibrines are common in fluids undergoing decomposition; they may be compared to minute chains, exceedingly flexible in their movements. Some species are not longer than the 300th of a line, and the 3000th of a line in thickness.

CONCRETIONS.

As it occasionally happens that the student, in the course of his investigations, meets with substances in various parts of the body differing in their nature from any of the elements hitherto described, it may be well to mention that a great variety of concretions are to be met with in the animal body, their nature varying with the locality in which they are found, their form depending upon the peculiar conditions in which they are placed. Thus, when a concretion occurs singly it is usually spherical, oval, or elongated; when numbers occur together they are cubical, pentagonal, hexagonal, or polygonal, according to the amount and direction of the attrition or pressure which they exert on each other during the movements of the viscera in which they are contained. Even foreign bodies, introduced into the internal cavities, undergo alterations of form from the operation of the same causes, as was illustrated some time ago in the case of a bullock whose rumen was found, after death, to contain fourteen large leaden bullets, every one of which had lost its spherical form, and become polygonal. Again, in the case of foreign substances that have accidentally found their way into the human bladder, portions of catheter, sealing wax, &c., are always found, after a time, to be moulded into peculiar shapes by the movements of the bladder.

Concretions occur in the liver, kidneys, bladder, salivary ducts, stomach, and intestines.

Biliary Calculi.—These are, perhaps, among the most common; they are usually of a greenish-yellow colour, but sometimes they may be of a dark yellow, brownish, or even black colour; occasionally specimens are met with nearly white. They vary in size from that of a pin's head to a hen's egg, the average being the size of an ordinary marble.

Biliary concretions consist, for the most part, of cholesterine, in different degrees of purity. The purer the specimen, the more crystalline will be the fracture, and the lighter the colour. They often show a concentric arrangement of layers, crossed by striæ running from the centre to the circumference, as seen in the drawing.

Fig. 168.

FRACTURED SURFACE OF A BILIARY CALCULUS.

Intestinal Concretions.—If we exclude the so-called hepatico-intestinal calculi, which are merely escaped gall-stones, intestinal concretions are very rare in man. They may be said to consist, for the most part, of imperfectly digested animal or vegetable food, or of a mixture of both. Sometimes they are composed of medicine, as happened on one occasion in a patient who passed a hard stone-like mass, looking like an uric acid calculus, but which, on analysis, was found to be gum benzoin encrusted with a small quantity of inorganic matter.

The patient had been taking pills of gum benzoin, in order to improve his voice, and these, instead of being digested and absorbed, appear to have accumulated in the intestines, and given rise to the concretion.

In the digestive canal of the horse and ox concretions are more frequent, consisting, generally, of triple phosphates arranged in concentric lamellæ round a central nucleus.

The nucleus is usually some foreign body, a small stone, a rusty nail, or an old button.

In some few instances a mass of undigested food has been found forming the nucleus of the concretion.

These inorganic intestinal calculi vary much in size, sometimes attaining an enormous bulk, being found 3, 4, 6, or even 8 inches in diameter, and weighing 6, 12, or 20 lbs. It is evident, from the history of many cases of animals in whose intestines these deposits occur, that very little inconvenience is felt from their presence so long as they remain in one position, but the act of shifting generally occasions colicky pains, and not unfrequently death results from the inflammation which follows.

The illustration conveys a good idea of a section of a small calculus from the intestines of a horse. The central nucleus and concentric layers are well shown.

Fig. 169.

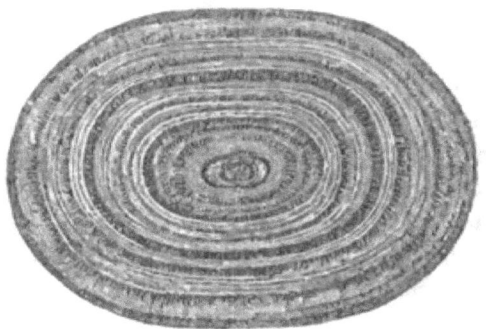

Section of a triple phosphate calculus.—Horse.

Animal Concretions may consist of imperfectly digested food, hair, and other animal matters. A singular specimen of the former was passed, after much suffering, by a gentleman aged 56. The mass, when fresh, measured nine inches in length and six and a half in circumference. Under the microscope it was found to consist of striated muscular fibres, cellular tissue, short portions of semi-digested bloodvessels, and a few hairs, the whole being held together by a quantity of mucus and lymph.

In another case large masses of fibrin, presenting, on section, a dense lamellated structure, were passed by the rectum.

Hair Concretions.—These are exceedingly common in oxen and other animals much given to licking themselves or their fellows. In shape the concretions are spherical, and appear to consist principally of the animal's own hair, as they resemble it in colour, and are usually of uniform tint throughout. After a time they become covered with a smooth crust of inorganic matter, which forms a shell, the interior consisting exclusively of roughly matted hair.

Hair balls are sometimes of an immense size.

In the human subject these concretions are but rarely met with; one specimen exists in University College Museum. It was taken from the stomach of an hysterical girl, who, after eating all her own hair, began to consume that with which the bottoms of the chairs were stuffed.

A microscopic examination is in general necessary to determine the nature of these concretions.

Vegetable Concretions.

Oat-hair Calculus is not unfrequent in horses, as might be concluded from the nature of their food; but few specimens have been obtained from the human subject.

These concretions are generally circular, often with a nodulated surface, having an external coat of mucus and inorganic salts, looking like a piece of brown sandstone. The section is of a pale coffee colour, and looks and feels like a piece of brown felt.

Under the microscope a number of hairs and vegetable spiral vessels will be observed.

The hairs are tubular and allow water to enter them by capillary attraction. In consequence of the air and water mixing in the tube, some of them present a curiously dotted central part.

The illustration is taken from an oat-hair calculus, one of twenty passed at different times by the same patient.

Fig. 170.

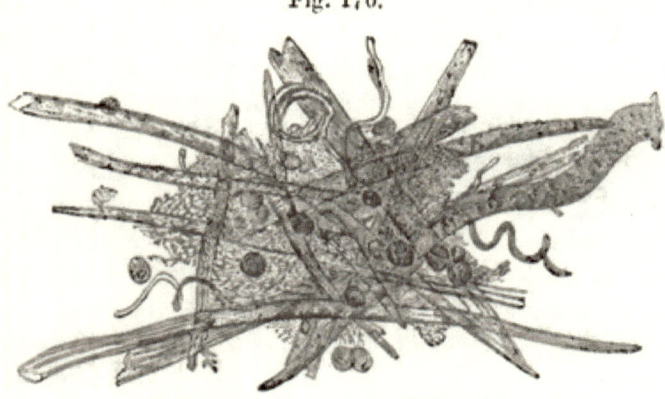

Oat-hair calculus.—Magnified 200 diameters.

Starch Concretion.—The patient from whom this mass was discharged had been labouring under dysentery. The mass, which was passed with some difficulty, consisted partly of ordinary fæcal matter, and partly of a hard white brittle mass of the size of a hen's egg. On being broken it resembled very closely a phosphatic calculus. Chemical and microscopic examinations, however, proved it to be composed entirely of hardened starch.

Urinary Concretions.—Urinary concretions are very frequent, both in the kidneys and bladder. They consist of the different urinary salts, as, for example, uric acid, urate of ammonia and soda, oxalate of lime, phosphate of lime, ammonio-magnesian phosphate, carbonate of lime, cystine, and xanthine.

Figures of the more common of these urinary deposits are given for the guidance of the student, under the head of 'Microscopic Examination of Morbid Fluids,' p. 182.

Besides the concretions mentioned there are a great variety of others to be met with in the animal body, for every glandular secretion is liable to yield them, but the foregoing examples are more than sufficient to make the student alive to the true nature of the products he may accidentally encounter; and for further information in any case of difficulty, he can always refer to some work on pathology, in which these matters are specially treated, our object being to give him a glimpse of the subject, to prevent him falling into errors in the course of his histological researches.

DEGENERATION OF TEXTURES.

A FEW words upon the degeneration of tissues are here absolutely necessary, otherwise the beginner may be frequently unnecessarily puzzled in the examination of different organs, by observing the normal constituents to be intermingled with, or even substituted by, products which are altogether different in character from the proper textures.

The process of substitution of diseased for healthy tissue is usually a slow one, except in the case of fatty infiltration, which often advances to an extreme stage in a short time. An important example of this is furnished by the occurrence of fatty degeneration of liver in Strasburg geese. These animals, being placed in a warm room, and frequently crammed with food containing an excess of fat elements, have their livers enormously enlarged and filled with fat globules. A few weeks, five or six at most, will suffice to convert a perfectly healthy liver into a mass of disease.

Muscular tissue is exceedingly prone to fatty degeneration, hence fatty heart is almost universal in domestic animals rendered obese by high feeding and little exercise, so much so that it is almost impossible to obtain a healthy heart or liver from any of the animals slaughtered for human food, in consequence of the present system of feeding. It becomes, therefore, requisite to select a heart or liver from a wild animal when a perfectly healthy example is required.

Fatty Disease of Muscular Tissue presents an appearance which is very characteristic. Under the microscope the muscular fibres will be seen to have partially or wholly lost their transverse striæ, while lines of fat globules are found extending along the fibres; sometimes they are the principal constituent.

The illustration is taken from a very advanced case of 'fatty heart,' from a gentleman who died suddenly while in the act of writing.

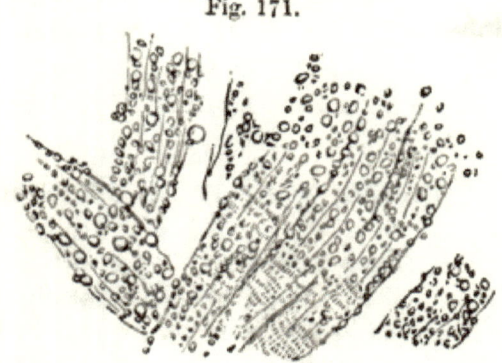

Fig. 171.

Fatty degeneration of the heart.—Magnified 200 diameters.

Fatty Degeneration of the Liver presents itself under various aspects, from the commencement of fatty deposit to the almost complete substitution of fat globules for the normal elements.

In the next illustration, the first drawing (A) represents the liver cells in the incipient stage of the disease.

In the second (B) the advanced stage of the disease is depicted, showing the degeneration and breaking up of the walls of the cells.

The third specimen (C), taken from a cat that had for years lived luxuriously, spending its time principally by the fire, is an example of complete substitution; hardly a trace of the natural textures of the liver could be discovered.

Fig. 172.

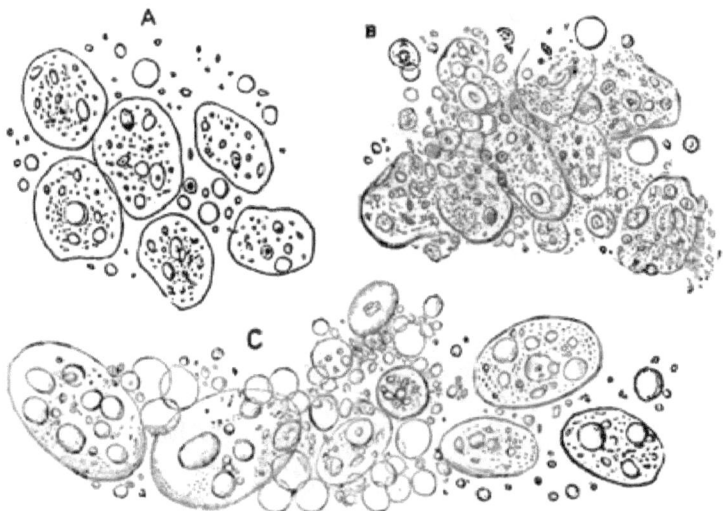

Fatty degeneration of liver. A. Incipient disease. B. Advanced disease.
C. Entire loss of normal structure.

Instead of an abnormal softening it happens, occasionally, that the opposite condition is induced by the deposit of mineral matters in the interstices of the tissue, constituting the change known as *calcareous degeneration.* It occurs in the bloodvessels, rendering them opaque and brittle, although in the early stage it appears as a mere whitening of the inner coat.

Calcareous degeneration of the muscles, cartilages, and ligaments is not uncommon. It occurs also in various morbid growths, especially in the walls of old hydatid cysts.

From the benign character of the deposited material it follows, that in many instances, calcareous degeneration is beneficial, when, for example, tubercular deposits, or cancerous growths, or growing fibrous tumours, or living parasitic organisms, become cemented into harmless and inert masses of mineral matter.

Tuberculous Deposit may be described as another form of degeneration, although it is more commonly included under morbid growths.

Almost all the internal organs are liable to the tuberculous change; muscular tissue, indeed, appears to be the only one capable of resisting it, hence it is we find tuberculous matter in the lungs, liver, kidneys, brain, spleen, lymphatic glands, and other secreting structures. The deposit is often small and circumscribed, but is varied in extent, from small grains of yellow or greyish colour (miliary tubercle) to considerable masses.

A typical specimen of tubercle contains numerous oval corpuscles, having minute granules in their interior, as shown in the illustration.

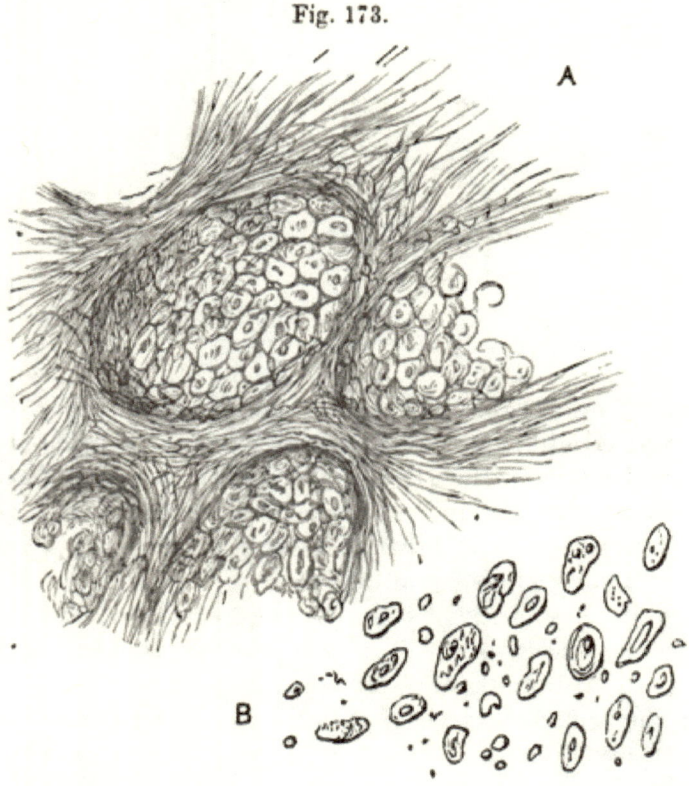

Fig. 173.

Tubercle. A. Deposit in the air-cells of the lungs. B. Miliary tubercle. Magnified 200 diameters.

Sometimes the exudation is surmounted by fibrous structure, constituting what is termed 'encysted tubercle.'

Under all circumstances tuberculous deposit implies a deficiency of vital energy in the nutritive functions, in consequence of which the process of cell formation is arrested at an early stage, and the deposited matter remains in the condition of granules, nuclei, or at best, imperfectly developed cells. The product has no malignant properties, and although it appears from a few cases that it may coexist with cancer, it must be distinctly understood that the two diseases have nothing in common.

In its physical characters tubercle differs considerably, being at one time hard, indurated, or calcareous, at another soft, cheesy, and even liquefied, in the latter case having the appearance of pus.

Indurated or calcareous tubercle is, on the other hand, principally composed of irregular masses of calcareous matter, the phosphate or carbonate of lime, with a few corpuscles, some granular matter, and sometimes crystals of cholesterine.

In liquid tubercle, again, the corpuscles closely resemble those of laudable pus, except that they are less regular in outline, and do not show the same well marked nuclei, after the addition of acetic acid. The cells, in fact, are more like those in putrid purulent matter.

Typhous matter, deposited during the ulcerative process in the small intestines, is of the nature of tubercle, consisting of minute granules with ill defined nuclei, without any indication of cell development.

Pigmentary degeneration is another form of disease, which it may be as well to allude to here; it essentially consists in the deposit of pigment granules in the tissues of an organ or part. At one time it was considered to indicate some degree of malignancy, but this cannot be the case, for it occasionally occurs in otherwise perfectly normal tissues. Thus we meet with it in the rete mucosum in cases of 'bronzed skin,' and even in the common freckles to which fair-complexioned individuals are so liable during the summer months. The idea of its malignancy doubtless arose in consequence of its occurring in malignant growths, in which it may in truth be considered as much an accidental product as it is in normal tissues.

Pigmentary deposits may occur in the crystalline lens, and give rise to what is termed *Black Cataract*.

A case of this somewhat unusual form of disease having been sent for examination gives us an opportunity of describing the general and microscopic appearances of the crystalline lens thus affected.

The external layers of the lens were sufficiently transparent to permit the dark central portion to be seen through them.

After peeling off the outer lamellæ, the central part was found to be of a dark mahogany colour, and exceedingly hard. On attempting to divide it by means of an ordinary dissecting needle, the instrument broke in two; but by using a stronger needle the nucleus was fractured, the broken surfaces being dry, white, and crumbling. Under the microscope the fibres of the dark portion of the lens were found to have colouring particles uniformly diffused in their substance. When treated with sulphocyanide of potassium after the addition of hydrochloric acid, they became red, and when further tested by ferrocyanide of potassium, they became blue, proving that the colouring matter was in great part iron or a salt of iron, and not melanine, which requires to be incinerated before it manifests the iron reaction.

The drawing of the fibres shows the arrangement of the colouring particles.

Fig. 174.

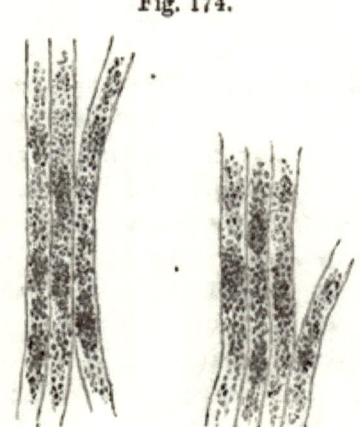

Black cataract. Fibres of lens with pigment granules.—Magnified 200 diameters.

Bronzed Skin.—In reference to this form of pigmentary deposit, it may be remarked that the integument is of a uniform dark hue, from the excess of colouring matter in the lower layer of the cuticle (*rete mucosum*), arranged in precisely the same manner as the normal pigment in a similar position in the skin of the negro and other black races of mankind.

The discovery of disease in the supra-renal capsules in certain cases of 'bronzed skin' formerly led to the idea that the disease of the capsule was the causes of the discoloration of the integument. The position, however, is at once disproved by the facts, that diseased supra-renal capsules are often found without 'bronzed skin,' and 'bronzed skin' with healthy supra-renal capsules. The coexistence of the two diseases has no more to do with their relation to each other as cause and effect than the simultaneous occurrence of any two or more dissimilar affections in different parts of the same organism.

The appearance of a section of bronzed skin under the microscope is indicated in the next drawing.

Fig. 175.

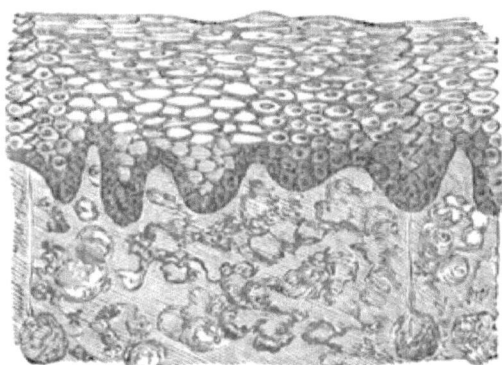

SECTION OF BRONZED SKIN.

Internal organs, as well as the cutaneous surface, are often the seats of pigmentary deposit. The cells of the liver, the peritoneum, and the pleura, both in man and the lower animals, are found thus affected. A few years ago, in the case of a horse, the lining of the thoracic cavity and all its contained viscera, lungs, heart, and vessels, were found to be covered with grape-like masses of black pigment.

Melanotic deposits are very frequent in grey horses, and, in old subjects particularly, some small tumours are generally present at the lower part of the root of the tail.

With reference to pigmentary degeneration of morbid growths, it may be stated that melanotic infiltration varies in different instances from a few black spots to absolute and universal blackness. This difference in degree may equally occur in a hard fibrous tumour and a soft encephaloid cancer.

The black matter deposited in all these cases is the same, viz. melanine, and just as in the case of the black cataract, iron is always abundantly present; indeed, iron is one of the constituents of melanine as it is of the colouring matter of the blood. In 100 parts of melanine there is one part of oxide of iron.

Melanine may exist either free or, what is more common, in the cells of a growth or tissue, giving to them a dark granular appearance.

From degeneration of textures we now pass on to a very brief consideration of adventitious growths.

ADVENTITIOUS PRODUCTS.

Osseous Tumours.

BONY growths occur in a variety of situations, in cellular tissue, the tendons and ligaments, cartilage, muscular fibre, the subserous tissue of the pleura, the submucous tissue of the gall bladder, and even the crystalline lens, are the seats of ossific deposit.

Osseous tumours are divided into Exostoses, Osteophytes, and Osteoid Tumours.

Exostosis is the growth of bone upon bone or its periosteum. A similar tumour growing into the medullary canal is termed enostosis. Exostoses differ in their density, sometimes being compact and firm, at others spongy, like cancellated tissue. Upon the limbs of horses exostoses are almost constant, either in the form of 'splents,' affecting the metacarpal bones, or 'spavins,' situated at the inner and lower part of the tarsus, besides being occasionally found in less common situations, as the head, lower jaw, ribs, and upon exposed parts of the skeleton generally.

While alluding to this animal it may not be out of place to mention a remarkable instance of disease of osseous tissue, which was sent to the laboratory for examination by Professor Varnell of the Royal Veterinary College.

The appearances presented by the morbid structures are depicted in the next illustration.

Fig. 176.

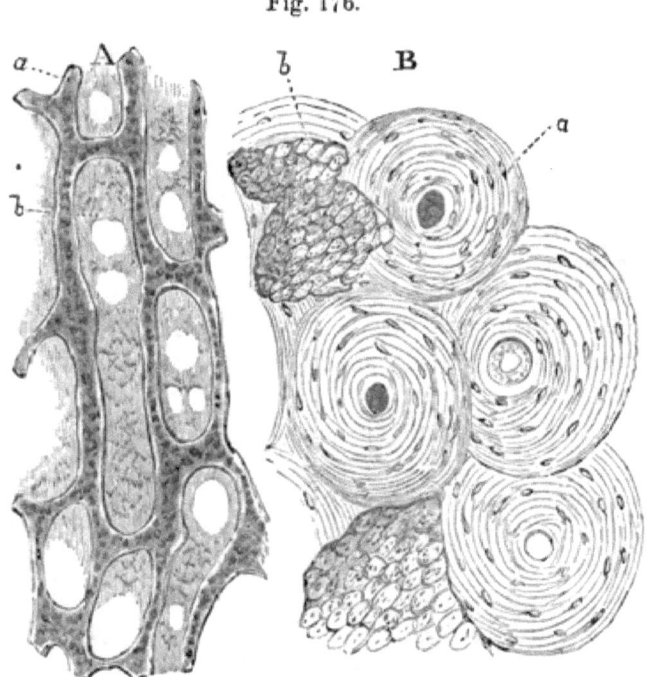

A. Longitudinal section of lower jaw. B. Section of one of the long bones.—Magnified 200 times.

In the letter accompanying the specimen it was stated that the disease broke out as an epidemic in a breeder's stud, and terminated in the death of nine or ten fine young horses. The first symptom observed was swelling and tenderness of one of the large joints. In a day or two these indications would subside or transfer themselves to the opposite joint. In a few more days several other joints would be similarly affected, and the animal be forced to remain still on account of the pain it suffered in the attempt to move. These apparently rheumatic symptoms were associated with hypertrophy and softening of the bones, and ended by the animals losing the power of walking, or even rising, in consequence, as was afterwards discovered, of the spontaneous separation of the ligaments from their attachments round the heads of the bones.

On post-mortem examination the viscera were found healthy, excepting that the muscles were pale in colour. On making an incision into one of the swollen joints, a considerable quantity of reddish-brown fluid escaped. The synovial membranes were, in some cases, thickened and vascular. The cartilages were ulcerated in patches varying in size from a pin's head to a shilling. At some points the cartilage was entirely absent and the bone exposed.

Under the microscope the bones presented two distinct kinds of morbid change, the cranial bones and ribs shewing alterations of structure entirely different from those found in the long bones.

The Haversian canals (*b*) in the section of the lower jaw are seen to be excessively developed, and the bony tissue (*a*) expanded into a reticulated structure. The osseous lamellæ are so diminished in number, that at the first glance they are likely to be mistaken for Haversian canals, and the canals themselves for bony tissue.

Under a high power the lacunæ were less distinct than in healthy bone; the canaliculi were in many places obliterated.

The long bones, being still dense and hard, could not be cut with a knife; accordingly a portion was immersed in dilute acid and a section made of the softened structure, which presented a very peculiar aspect. In general the Haversian canals with their lamellæ appeared normal; the lacunæ and canaliculi, as far as could be seen in the decalcified bone, were healthy; but, here and there irregularly shaped cavities were seen in the substance of the osseous tissue, filled with fat cells and granules, as shown in the drawing, fig. 176, B.

The only approach to this condition is observed in *Mollites ossium*, but in that disease the fatty matter is deposited in the dilated Haversian canals.

Osteophytes are bony growths developed under the influence of one of the bones from an extra-osseous exudation. They are often very slightly attached to the adjacent bone, and may be readily separated from it, although they are sometimes closely cemented and inseparable. In form they vary considerably, occasionally assuming very fantastic shapes. They may be flat, triangular, semicircular, foliaceous, stalactiform, cauliflower-shaped, or stellate; occasionally they occur as membraniform plates between the cranium and dura mater in pregnant women.

A specimen of the foliaceous form of osteophyte is shown in the illustration.

Fig. 177.

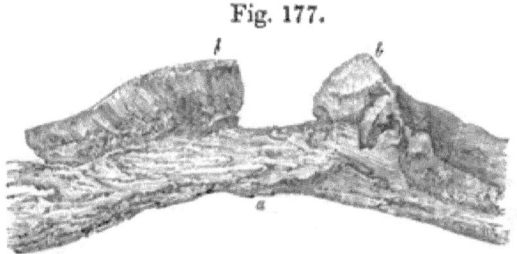

Foliaceous osteophyte of the clavicle; the foliæ (*b b*) running at right angles with the axis of the bone (*a*). U. C. Mus.

Osteoma.

Under this name is described a species of osseous growth of either quick or slow development, generally arising from the surface of a bone, composed of dense or porous osseous tissue, with a quantity of greyish white vascular substance of the consistence of fibro-cartilage lying in its interstices. The soft substance, examined under the microscope, is found to be constituted of a fibrous network inclosing nucleated cells in its meshes. These tumours sometimes attain considerable bulk.

From the circumstance of one tumour being occasionally followed by others, even in the soft parts, including the lungs (whether the original tumour has been removed or not), it is supposed that their formation is due to some constitutional cause.

The idea has also been advanced that osteoma is a cancerous growth in which the stroma has undergone ossification.

Osteosarcoma is a term applied to encephaloid growths arising from bones, and receiving offshoots of bony tissue into their interior.

The bony processes are stalactite in form or radiate, resembling in some instances the arrangement seen in osteoma.

Enchondroma.

Growths simulating cartilaginous structure occur most frequently upon bones. The metacarpus and phalanges appear to be the favourite seats; the tumours have been met with in the cranium, upon the ribs, and in glandular structure.

Enchondromatous growths are generally globular, with a smooth surface, or otherwise somewhat tuberous. When cut into they are found to be divided into small irregularly rounded cavities or loculi, which are filled with a gelatinous substance of a greyish or greenish yellow colour, the entire aspect being not unlike colloid, with which disease enchondroma may be confounded in a mere eye inspection.

Under the microscope the walls of the loculi are found to consist of fibrous tissue, or sometimes of a substance resembling the matrix of true cartilage. The gelatinous matter consists of cells sometimes granular, but more generally containing nuclei or secondary cells in their interior.

The presence of bony matter in the interior of an enchondromatous growth is not uncommon, and occasionally the walls of the loculi undergo the ossific change.

The illustration represents a section of one of these tumours as it appears to the naked eye.

Fig. 178.

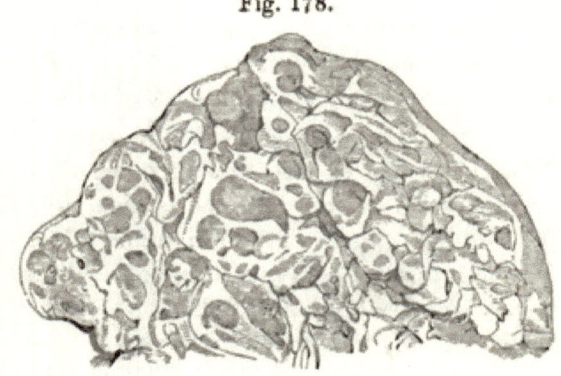

Section of enchondroma, showing the numerous loculi filled with gelatinous matter.

Fibrous Tumours.

These products occur in various situations, either singly or in large numbers. The interior of the uterus is their most usual seat.

Sometimes fibrous growths are closely connected with the surrounding parts, at others they are almost free, being only attached by a pedicle, while occasionally they are contained in a species of cyst, with which they have no connection save that of contact.

In form they vary but slightly, being generally spherical or presenting some modification of a sphere.

Fibrous tumours are little prone to change. It occasionally happens, however, that they are attacked with inflammation, in consequence of which they ultimately soften; on the other hand, they may remain for a long period and finally undergo calcareous degeneration.

Recurrent Fibroid is the name given to a variety of fibrous growth that manifests a disposition to return, each succeeding tumour approaching nearer to the character of a malignant formation. Recurrent fibroid does not differ in aspect or microscopic characters from simple fibrous tumours, which seem to pass, by almost imperceptible gradations, into the malignant class.

Fibrous growths, when cut, present a glistening appearance, the surface resembling a watered ribbon. Generally they are devoid of juice.

Cysts filled with a jelly-like matter resembling colloid sometimes occur in the substance of the tumours.

Melanotic deposit is not unfrequent, particularly in the fibrous growths appearing in cattle at an early age. The deposit is sometimes sufficiently abundant to render the tumour quite black; it does not, however, seem to indicate any malignant tendency, as such tumours manifest no disposition to return after removal.

Fibro-fatty tumours contain fat vesicles more or less abundantly intermixed with the fibrous structure.

A portion of a fibrous tumour, examined under a high power, after being well teazed out, will be found to present certain differences of elementary structure according to the degree of development the growth has attained.

The more recent deposits consists of a blastema, in which numerous nuclei are imbedded. A higher degree of advancement is indicated by the presence of elongating fusiform or fibre cells. The most perfect growths are composed of fibres identical with those of connective tissue, intermingled with yellow elastic fibres.

Acetic acid should be employed in this examination for the

purpose of rendering the nuclei evident, and also to enable the observer to detect yellow fibres when present.

The illustration shows the elements of an ordinary fibrous growth from the human uterus.

Fig. 179.

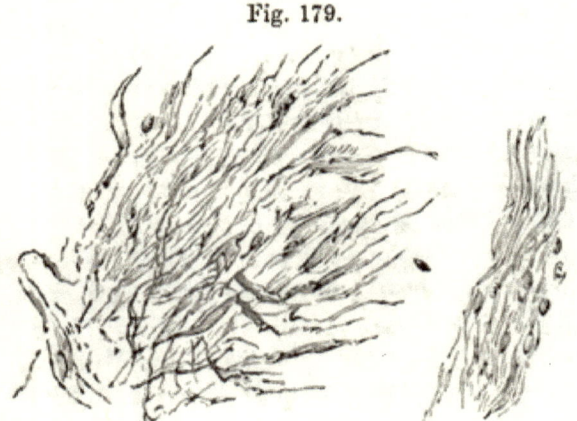

Fibrous tumour.—Magnified 200 diameters.

Indurated Chancre, although due to a specific infection before ulceration, appears under the microscope to consist of elements almost identical with those of some forms of fibroid tumours. The indurated deposit is intimately connected with the substance of the true skin by a number of white fibrous cords, as if an infiltration of plastic matter had taken place in the meshes of the cutis vera.

The indurated structure itself contains the developing elements of fibrous tissue, viz. nuclei, round and oval, with fusiform fibre cells, advancing to the condition of the true fibre of connective tissue. In one stage of its development, when the period of incubation has been long, and the infected part feels hard and circumscribed, like a split pea under the skin, numerous cells, identical with those of fibro-plastic growths, are to be found in it. After ulceration begins these gradually break up, and ultimately disappear.

Fatty Tumours (Lipomata).

These growths most frequently occur in the subcutaneous cellular tissue in various parts of the body. Upon the scalp and back they are sometimes very numerous. Occasionally they are met with in the lungs, liver, kidneys, bones, and even in joints.

They are usually globular in shape, with a smooth or slightly lobulated surface. When they occur in joints they are commonly branched, and receive the name of *Lipomata arborescentia*.

In size they vary to an extraordinary extent, sometimes being no larger than a pea, at others reaching the enormous weight of 40 lbs.

Common fatty tumours are composed of fat vesicles, identical with those of healthy adipose tissue, closely packed, and held together by a variable amount of connective tissue.

The structure is best seen by placing a thin section, cut with a razor, on a glass slide in a little water, and carefully covering it with a piece of thin glass, without employing any pressure, which would cause the rupture of the vesicles, and spoil the preparation. Under the high power the student will recognise the structure of ordinary adipose tissue, as represented in fig. 40, p. 55. Certain modification in the arrangements of the elements are occasionally met with in steatoma, cholesteatoma, and recurrent lipoma.

Steatoma is a variety of fatty tumour consisting of amorphous granular fat, without any true fat vesicles. In texture these growths are close-grained and inelastic.

Cholesteatoma is another form of fatty growth, analogous to the last, but containing, in addition, numerous crystals of cholesterine. It has a somewhat pearly aspect, or dull white colour, with a lamellated arrangement.

Recurrent fatty tumours are distinguished by a tendency to return after removal. They seem occasionally to possess some of the elements of malignant growths, and to have been even mistaken for them, as the two following cases will show.

The first instance is that of a morbid growth removed by operation from a woman's leg after having thrice recurred. The tumour, on section, presented a somewhat glistening aspect, was white in colour, friable and soft in texture. From its appearance to the naked eye, and from the fact of its having returned for the third time, the surgeon came to the conclusion that it was an example of encephaloid.

Under the microscope, however, the tumour proved to be a good specimen of an ordinary fatty growth.

It consisted of large fat vesicles, with only a few fine connecting fibres, and here and there one or two granular cells scattered among them.

The second specimen was removed from the sole of a child's foot. The first growth had been taken off when the child was a few weeks old, and very shortly afterwards a second made its appearance and grew with rapidity. The tumour was firm and elastic to the touch, had two or three hard nodules on its surface, with here and there a bluish pink tinge.

The history of the case, coupled with the appearance of the growth, led the surgeon to suppose it to be malignant, and accordingly he amputated the foot.

The illustration represents the reticulated appearance of the section.

Fig. 180.

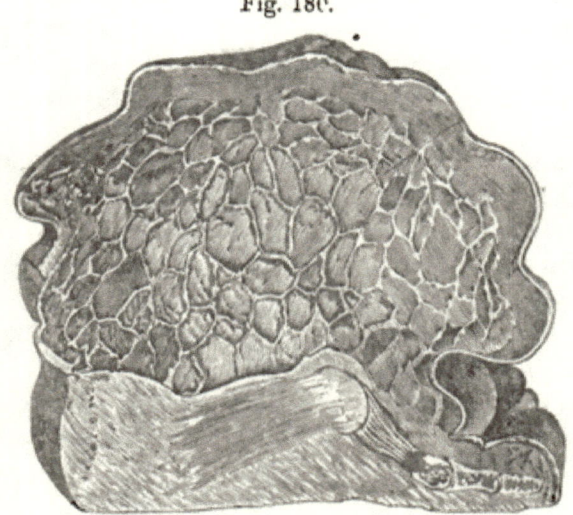

SECTION OF RETICULATED FATTY TUMOUR ON A CHILD'S FOOT.

Internally the tumour was found to be yellowish in colour, with a fine white reticulated stroma intersecting it in all directions, the yellow substance occupying the meshes or loculi formed by the white stroma. Under the microscope the stroma was found to consist of fine fibres of areolar tissue, and the yellow substance to be fat vesicles, many of which contained crystals of margaric acid arranged in characteristic stellate groups of fine needles.

The conclusion drawn from the microscopic investigation was that the tumour was a fatty growth containing an unusual quantity of areolar tissue. From its peculiar aspect the term 'reticulated fatty tumour' was applied to it.*

Myeloid Growths.

Myeloid Tumours are so termed in consequence of the resemblance of their elements to the marrow cells of young bones. They are composed of large oval or irregularly shaped corpuscles, sometimes occurring in masses and containing a variable number of nuclei, from two or three to ten or twelve; free nuclei, fibre cells, and fatty granules, with a quantity of granular matter, are also present.

Myeloid growths are generally found upon or in bones. In form they are oval or spherical, rather compact and firm, but brittle. The cut surface is of a greenish or greyish white colour, variegated with occasional crimson patches.

In one instance a myeloid growth upon the frontal bone of a young child led to a fatal result, from the supervention of meningitis. The morbid mass in this instance was wanting in firmness, being rather gelatinous, and was readily washed away from its attachments with water.

Hæmatoma.

Vascular growths are not of frequent occurrence, and when present seem mostly to consist of dilated vessels, both veins and arteries. In a few cases of brain disease in sheep and horses, these vascular collections have been discovered at the base of the cerebellum.

The term *Angeiectoma* is applied to those enlargements which consist of dilated vessels. Another variety, however, occurs most commonly in superficial parts, the skin, and cellular membrane, and is distinguished by the structure and physiological properties of erectile tissue.

---

* This important case was published in the Pathological Society's Transactions.

A section of one of these growths is represented in the drawing.

Fig. 181.

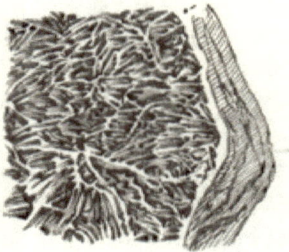

Section of an erectile tumour, showing its cavernous structure.

The term *Hæmatoma*, or blood tumour, is, however, applied to tumours which are non-vascular, consisting originally of effused blood, and ultimately of a mass of fibrin devoid of colour, the colouring matter having been removed by absorption.

Such tumours are sometimes filled with fluid contained in numerous cysts or loculi.

Under the microscope these growths exhibit the ordinary characters of fibrin, viz. minute fibrils which are gelatinised by a drop of acetic acid, amorphous matter, and granules.

EPIDERMOID AND EPITHELIAL GROWTHS.

The excrescences which are commonly found projecting from the surface of the skin or mucous membranes do not consist exclusively of accumulations of epithelial cells; on the contrary, it will be found that all the tissues of the part are implicated in greater or less degree in the morbid changes that have taken place. Thus, in dermoid growths, the epithelial layer, the papillated structure, and the cutis vera are all in a state of hypertrophy.

Keloid tumours are examples of that form of disease where the cutis vera is most affected.

Elephantiasis is an instance of the cutis and cuticle being about equally affected.

Ichthyosis is principally confined to the epidermis.

The next drawing includes specimens of dermoid growths taken from 'foot-rot' in sheep, 'canker' in the foot of the horse, and also from the disease of 'thrush,' affecting the frogs of the horse. A comparison of the three forms will show how closely allied are the products of these different diseases.

Fig. 182.

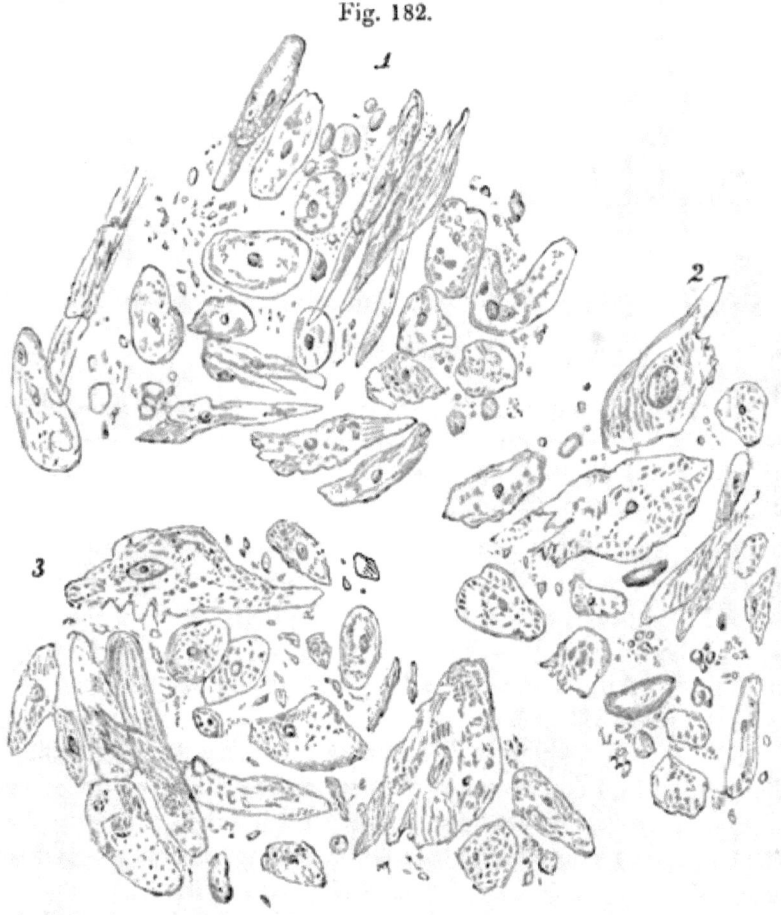

1. Portion of matter scraped from the cleft of the frog of a horse affected with chronic 'thrush.' 2. Fungoid matter from a sheep's foot affected with 'foot-rot' in an advanced stage. 3. Scraping from the fungoid surface of a horse's foot affected with inveterate 'canker.'

The most simple forms of dermoid growths are those which arise from the secretion of a soft apparently fungoid material in the place of the normally dense epithelial covering, as in cases of 'foot-rot' in sheep, 'canker' and 'thrush' in the foot of the horse, in which diseases the introduction of irritating particles, such as sand or dirt, through fissures in the hoof, occasions disease of the secreting structures, leading to elongation of the papillæ and the production of fungoid excrescences which spring up from the surface with extraordinary rapidity.

Under the microscope they are found to consist of the epidermoid cells of which the hoof—being, like the human nail, modified epidermis—is composed. Sometimes the cells are elongated, and occasionally their edges are somewhat ragged, especially in old cases, but they differ in no important particulars from the elements of the normal hoof.

Peculiar forms of Disease affecting the Nails of the Human Subject.

The two following cases of disease of these cutaneous appendages are quoted as possessing considerable interest, and serving to illustrate the identity of morbid changes occurring in similar tissues in man and the lower animals.

In the first case a gentleman's thumb-nail gradually became elevated by the deposit of a soft mass of epithelial matter, which pushed the normal structure from its position, much in the same way as the fungoid growths in 'foot-rot' of sheep, and 'canker' of the foot of the horse cause the detachment of the hoof-horn from the parts affected.

Medical men frequently denominate these affections psoriasis of the nail, and no doubt many of them deserve the title, but others possess more the characters of epithelial warts than anything else.

The present case was of this latter variety, and was cured by repeated scrapings and applications of nitrate of silver.

In the accompanying woodcut a representation is given of the diseased deposit beneath the nail, in juxtaposition with the next case, which presented features of a totally distinct character, for instead of there being hypernutrition, causing excessive development of epithelial cells, there was exactly the reverse condition, viz. defective secretion, with detachment of the nail from its matrix.

In the illustration three forms of disease affecting nail and horn are depicted.

Fig. 183.

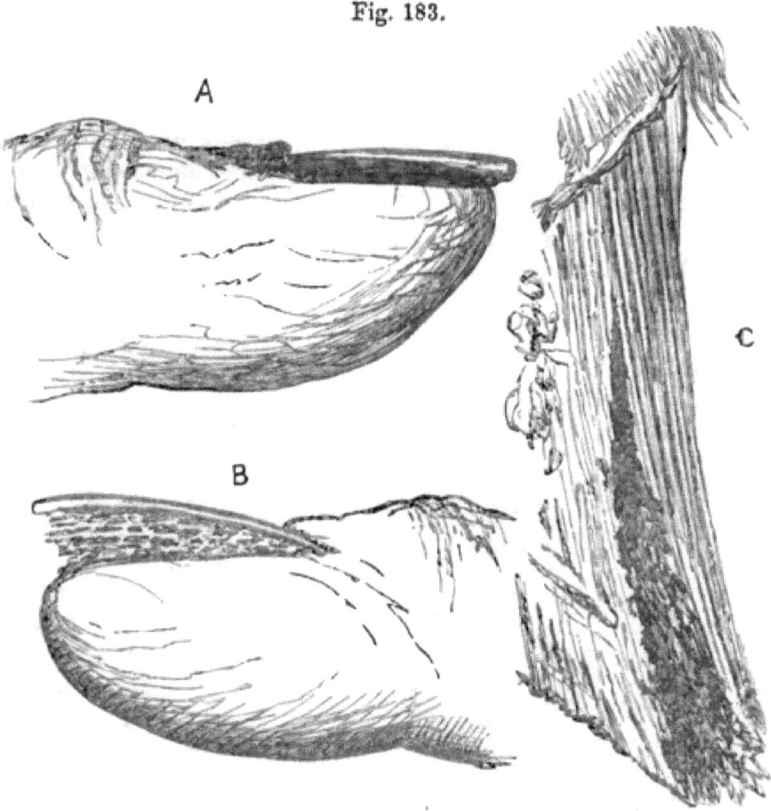

A. Separation of the nail from its matrix; the black line shows the extent to which the dirt has reached. B. Dermoid growth beneath the thumbnail, pushing the nail from its situation. C. Section of horse's hoof, affected with 'seedy toe,' showing separation of the layers of cells forming the hoof, and the accumulation of dirt in the space.

'A in the figure refers to the case of a woman aged 33 years who suffered from general debility, and among other phenomena, a peculiar disease affected the nails, which became separated from their matrix, allowing the dirt to pass by degrees nearly the whole extent of their length. The hair at the same time came off in abundance. Ultimately the nails were so far disconnected from their attachments that it was necessary to use finger-stalls or strips of adhesive plaster to retain them in their situations. After long-continued treatment, directed to the improvement of the general health, the case may be said to have terminated successfully, although there is still a greater amount of detachment than exists in the perfectly normal nail.'

The disease, in its main features, bears a remarkable analogy to the affection known as 'seedy toe' in the foot of the horse, a similar separation between the layers of the horn occurring in this disease, associated with the introduction of dirt, which aggravates the evil and often leads to continued lameness.

Elephantiasis affects both the cutis vera and epidermis, and is best studied in longitudinal and transverse sections, which will show all the structures to be in a condition of hypertrophy.

Fig. 184.

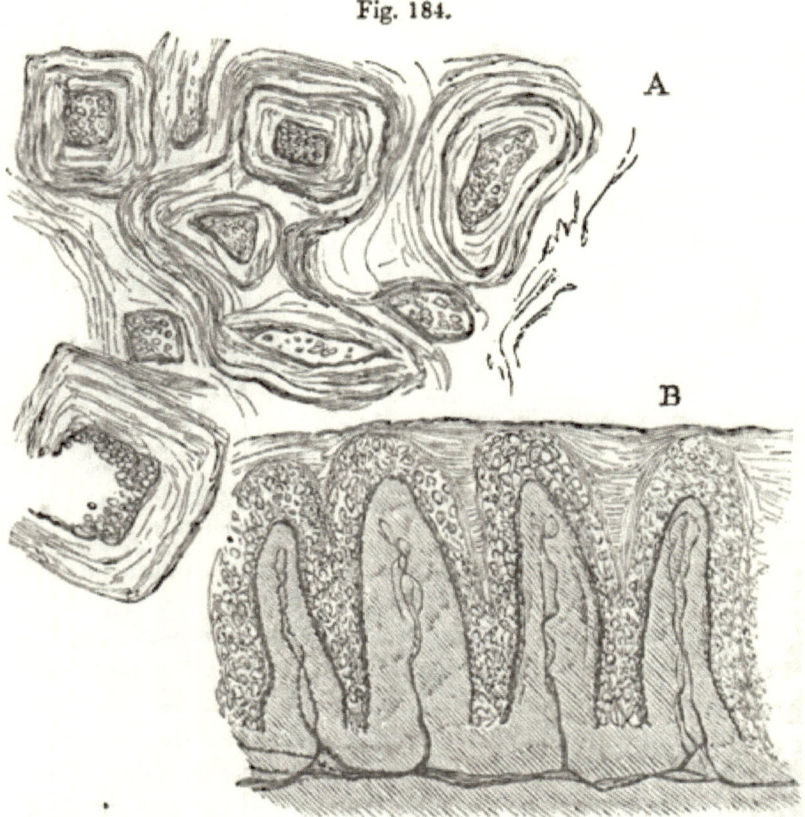

Sections of integument in elephantiasis. A. Transverse section near the surface, showing the enlarged papillæ surrounded by layers of epithelial cells. B. Vertical section through the skin, showing the elongated papillæ.

The papillæ of the true skin are much enlarged and elongated (as shown in fig. 184 B, which represents a longitudinal section) and round them layer upon layer of epithelial cells are secreted and arranged, exactly as in the healthy skin, but in an exaggerated form, very much resembling the structure of the hoof of the horse, save that the concentric lamellæ are less regular in outline (fig. 184 A).

Epithelial Growths, occurring upon the cheeks and lips, and especially upon the penis and scrotum of chimney-sweeps, have been classed among malignant tumours, under the name of *Epithelial Cancer*, in consequence of their extending to the deeper seated structures, and sometimes undergoing the process of softening or ulceration.

The illustration represents two typical specimens, one (A) taken from the sciatic nerve, and the other (B) from an 'epithelial cancer' of the leg attended with ulceration.

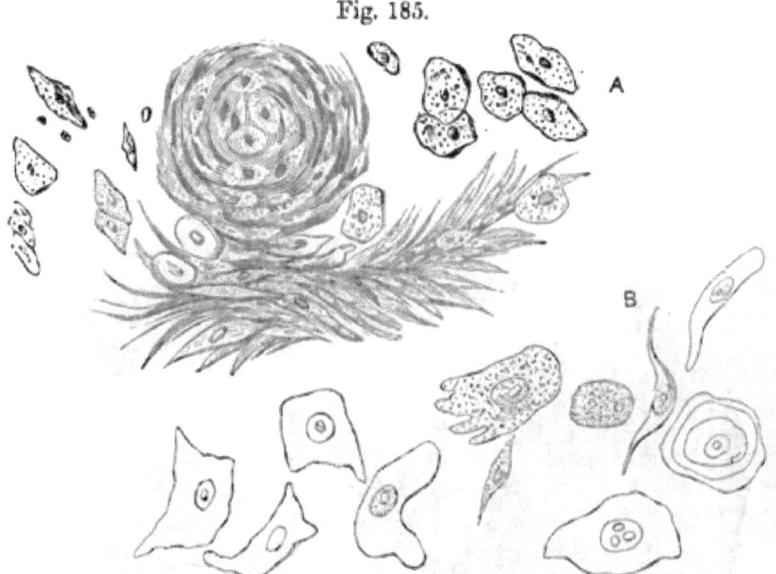

Fig. 185.

Epithelial cancer. A. From the sciatic nerve. B. Cells from an epithelial growth upon the leg.

Like other tumours they may degenerate and become *cancerous*, but in their ordinary form they do not appear to merit the title of *malignant*, as they neither produce the 'cancerous cachexia' nor affect the glands, and, when completely removed, they are not liable to return.

These products appear in the form of tolerably well-defined hard nodulated tumours, covered with minute warty papillæ. When situated upon a mucous membrane they often assume the form of cauliflower excrescences.

'Epithelial cancer' has been found in the liver and lymphatic glands, and even in nerves.

Under the microscope the tumours or deposits prove to be composed of epithelial cells, many of which are arranged round a central large cell containing new cells in its interior, apparently being developed by the endogenous process.

Adenoid.

Under this term is indicated hypertrophy of glandular structure. The human mammary gland is especially liable to this condition; and many of the so-called fibrous, and even malignant, tumours of this organ amount to no more than simple hypertrophy of its glandular elements.

The drawing represents a fragment of a diseased mammary gland after being slightly teazed out and subjected to pressure.

Fig. 186.

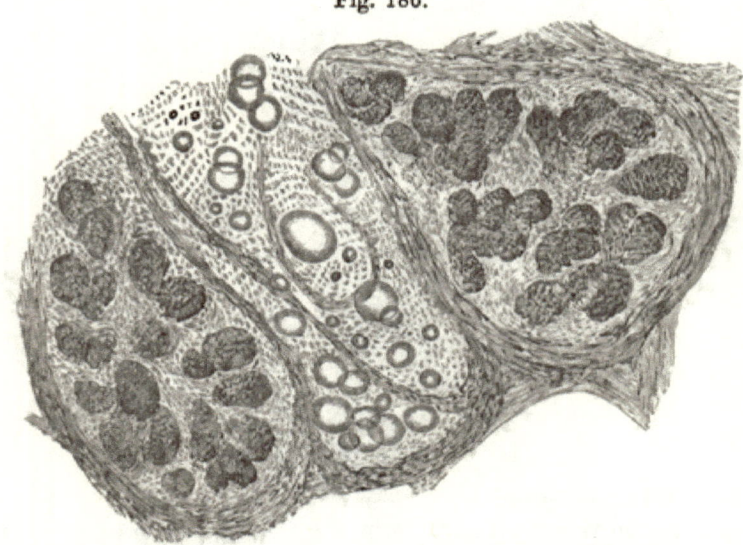

Adenoid tumour from mammary gland, showing the hypertrophied vesicles.

In these growths the characteristic ducts and lobules, with the epithelial cells of the normal gland, are readily detected with the aid of the microscope.

A considerable increase of the connective tissue gives to the hypertrophied portion more than the usual firmness of the normal gland structure.

Cystoid Tumours.

Cystoid growths may consist either of an enlargement of an already existing cavity, such as a sebaceous follicle or synovial sac; or they may occur as new productions, either singly or in groups.

Sebaceous Cysts are very common on the scalp, from the obliteration or diminution of the orifices of the sebaceous follicles causing the secretion to accumulate. The contents of cystic tumours differ in appearance, sometimes looking like gruel (atheroma), or water (hygroma), or honey (meliceris). Under the microscope the constituents are found to be fatty granules, epithelium, cholesterine tablets, and hairs. The cysts are lined by a layer of epithelial cells.

Similar tumours occur in the eyelids, the lips, mouth, and uterus, from obstructions to the openings of the Meibomian gland and mucous follicles.

Glandular Cysts may arise from primary distension of the vesicles of the gland.

Synovial Cysts result from distension of the synovial bursæ, which are found in the vicinity of joints, by excessive secretion of synovial fluid. These sacs, under the name of 'windgalls,' are very common in the extremities of the horse, particularly in the neighbourhood of the fetlock joints.

Cysts originating independently of previously existing cavities occur in the cellular tissue, from pressure or friction, and also in the glandular structures, or rather in the fibrous tissues investing them; they may also happen in tumours, whether benign or malignant.

Some of the domesticated animals present frequent examples of serous cysts, originating from injury.

In horses we have them arising in consequence of undue pressure from a part of the harness, generally upon the back or sides of the 'withers.'

In dogs the affection known as 'tumefied flap of the ear' is a remarkable example of the production of a perfect cyst quite independent of any previously existing cavity. The disease only occurs in dogs with pendulous ears, as setters and spaniels, and manifests itself by swelling on the inner surface of the 'flap,' consequent upon the effusion of fluid into the subcutaneous cellular tissue of the part. In many cases the cavity formed is found to be divided into numerous compartments, either by extension of the fibres of the normal connective tissue, or by a new fibrinous deposit, so that, in order to evacuate the contents of the sac it becomes necessary, after opening it, to cut or break down the membraneous partitions which hold the fluid as it were in separate cells.

Compound Cysts are the result of the development of numerous smaller cysts from the walls of the parent, forming a multilocular mass.

Fig. 187.

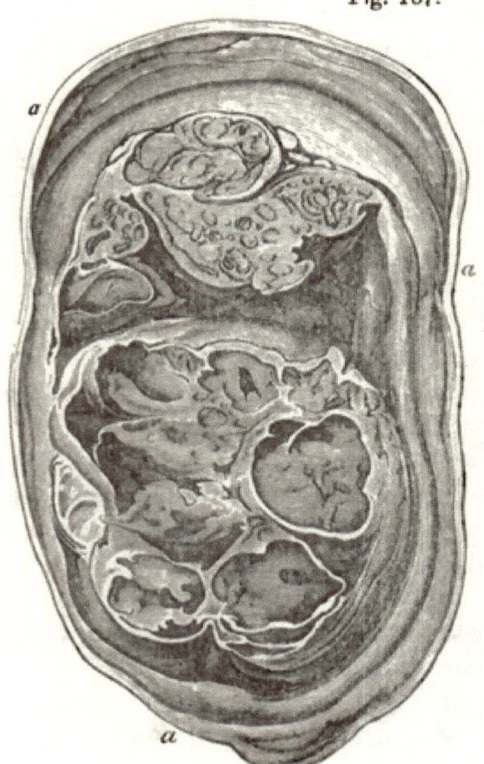

Compound or proliferous ovarian cyst (ad nat.). *a, a, a*. Divided walls of the principal single cyst. *b*. Small simple cyst. *c, c*. Two masses of compound secondary cysts, containing many of a tertiary order.

The foregoing illustration (fig. 187) represents a section of such a tumour from the human ovarium.

It may be remarked, in passing, that many so-called cystic tumours will be found, on examination, to be the caudal vesicles of hydatids.

Colloid.

Colloid tumours are characterised by an alveolar arrangement of the internal structures. The walls of the alveoli are composed of fibrous tissue and are sometimes extremely delicate; in other parts they are tolerably well developed.

Some of the loculi are of the size of a grain of sand, others are much larger, and contain a jelly-like substance of a greenish yellow colour, consisting of caudate, fusiform, and granular cells, both nucleated and non-nucleated. Some of the cells contain others in their interior, endogenously developed.

The drawing indicates the general appearance of a colloid tumour in the ovary.

Fig. 188.

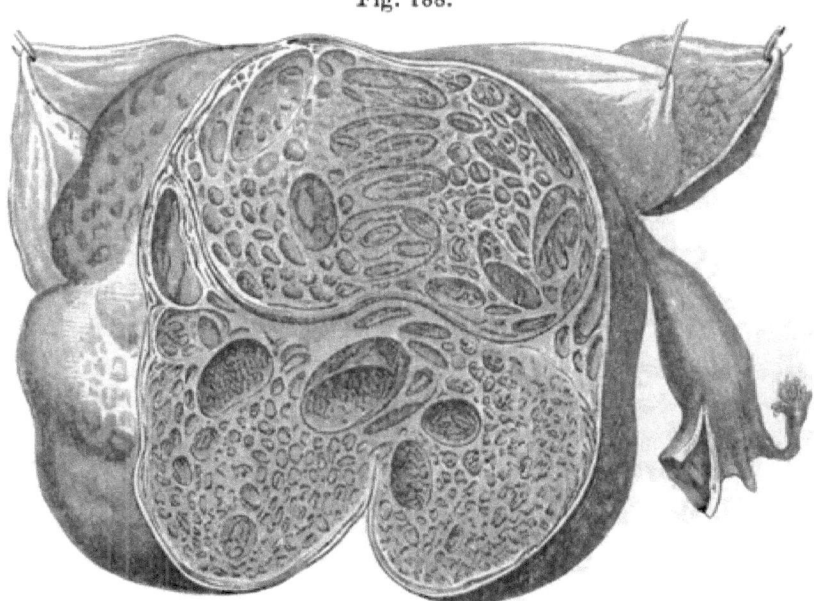

Colloid cancer of the ovary.—After Cruveilhier.

Colloid occurs in the stomach, omentum, kidneys, uterus, and spleen, and also in the ovarium. Its growth is often rapid, but it does not affect the glandular system nor produce 'cancerous cachexia,' hence it cannot properly be classed among malignant products.

Schirrus.

Schirrous growths possess four chief characters:—
1. They are hard. 2. They have a well-defined outline. 3. The aspect of the cut surface is glistening. 4. There is no juice.

The tumours, which are usually of slow growth, consist of a blastema, in a more or less advanced state of fibrous development, inclosing a variety of cells, fusiform, spheroidal, and irregular, having nuclei or fat globules and granules in their interior.

The most common situations of schirrous growths are the mammary glands, the pylorus, and the rectum.

The illustration is from a schirrous tumour of the human mammary gland.

Fig. 189.

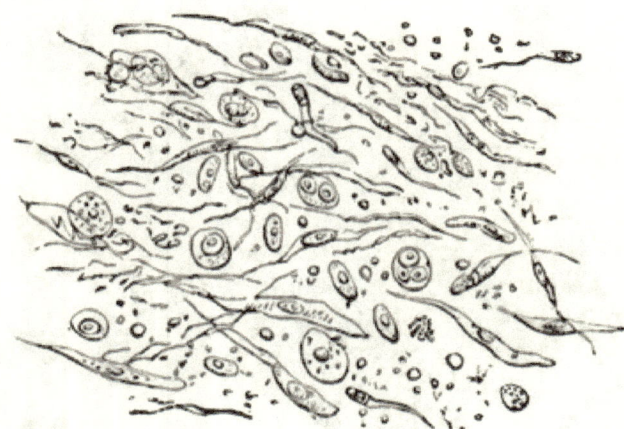

Schirrous growth from mammary gland.—Magnified 200 diameters.

Schirrous growths, on boiling, yield gelatine, a little albumen, and a considerable quantity of saline matter.

In true schirrus there is no affection of the glands; indeed, the 'cancerous cachexia' is entirely absent. It is only when the tumour begins to degenerate into encephaloid that symptoms of malignant disease present themselves in the patient.

The simple schirrous tumour, when fairly removed, is not apt to recur, which is a reason the more for not placing it in the same category as encephaloid, or regarding it in the light of a malignant tumour.

It would be much better for the histologist as well as for the surgeon, were the term *cancer* to be entirely abandoned, or at least limited to those tumours which have a constitutional effect, infiltrating the glands and poisoning the blood to such a degree that their removal is attended with no ultimate benefit to the patient, as they are certain to recur, and that, too often, in a worse form. At the present moment the term *cancer* is applied to a multitude of growths which do not possess the slightest similarity; for example, to schirrus, colloid, epithelioma, and encephaloid, not one of which has, either to the naked eye or microscope, a single character in common.

The name *cancer*, or malignant growth, ought only to be given to a tumour which induces that peculiar state of body distinguished as *cancerous cachexia*, infiltrates the neighbouring glands, and is certain to return, even after complete removal, either in the same or in some other situation.

The mere fact of a tumour recurring in the same situation after removal is in itself no evidence whatever of malignancy, as this may and does happen repeatedly in the cases of benign growths. The difference, however, is that while the malignant recur, no matter how complete their removal has been, and the second is, as a rule, in a worse form than the primary growth, the benign only recur when a fragment, no matter how minute, of the original growth has been left after the operation. This is now and again seen in simple schirrus of the female breast, where little offshoots are frequently overlooked, and these, after a time, become visible tumours; but even then, if the tumour has been a simple schirrus, the new growth ought not to be called secondary, seeing that it is only an increase of a part of the original tumour, still retaining the same benign character, and does not cause the neighbouring glands to become affected, which would inevitably be the case were it of a malignant or semi-malignant type.

Encephaloid. Cancer.

In pointing out the characters of a malignant tumour it will be best to take a typical specimen, that is to say, one that presents the special features of the class in a marked degree. It must be borne in mind that all tumours have a tendency to run into each other, the characteristic features of each becoming gradually less and less distinct, until the malignant ultimately appear to be almost identical with benign, and the benign with malignant growths.

Microscopists of the present day have abandoned the fiction of a special *cancer cell*, to the existence of which so much importance was formerly attached, as being the essential element of malignant tumours.

In many instances the naked eye characters of the morbid product are so decided that no difficulty exists in determining its malignant nature; but in other cases, not only is it necessary to obtain the aid of the microscope, but, in addition, all the evidence afforded by the locality of the tumour, its effect upon adjacent tissues, and also upon the patient's constitution, before a correct opinion can be formed.

Encephaloid bears a close resemblance to brain tissue, and hence its name. It occurs in various situations: the brain, lungs, liver, kidneys, spleen, and lymphatic glands are its most common seats. According to its aspect and physical characters it receives the name of 'mastoid' (like the boiled udder of the cow), 'solanoid' (potato-like), 'nephroid' (like the section of a kidney), 'hæmatoid' (when unusually vascular), 'melanoid' (when it contains patches of pigment in its structure).

The terms 'medullary sarcoma' and 'medullary fungus' may be considered identical with encephaloid.

In consistence encephaloid tumours vary considerably, some being comparatively firm, others almost semifluid, but they all have more or less of a milky juice, which, even of itself, is almost sufficient to distinguish them from all benign growths.

The juice, when pressed from the tumour, is creamy or albuminous looking, and when examined under the microscope is found to contain the same cell elements as the tumour, which is principally composed of cells of various forms and sizes, spherical, caudate, and polygonal cells, with or without large nuclei; also large granular cells, some containing fat granules and others a quantity of dark pigment, large cells developing others within them endogenously; free granules and nuclei; and, lastly, a small amount of fibrous stroma, serving to connect these elements together.

The illustration includes the most characteristic elements of encephaloid growths.

Fig. 190.

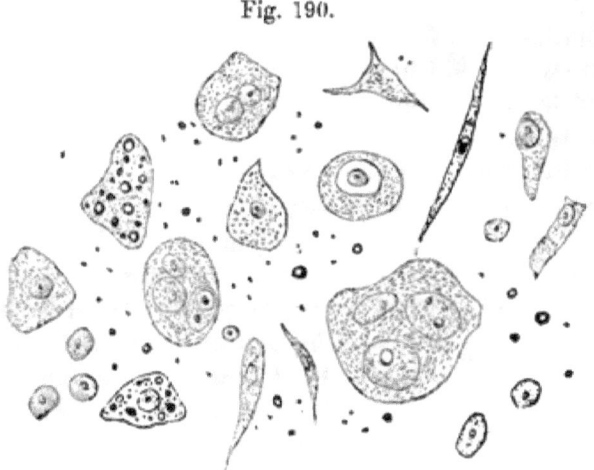

Elements of an encephaloid tumour.—Magnified 200 diameters.

Malignant disease may occur in the form of a single tumour, but it is much more common for several to coexist, either in the same or in different organs, and this is one of the reasons why the removal of an encephaloid growth is almost certain to be followed by its return. By some it is considered that the very act of interference rouses into activity a previously almost dormant developmental force, inducing products of even greater malignancy than the original growth.

As it is of the utmost importance to be able to say what is and what is not malignant, it may be convenient to recollect the following characters, which, when coexistent, at once decide the growth to be malignant in its nature.

1. The tumour is brain-like (mottled red and white).

2. It has a milky juice, which, under the microscope, is found to be full of various cells and nuclei.

3. The growth has no well-defined outlines, but gradually shades into the surrounding parts.

4. It is generally soft and pulpy.

5. The neighbouring lymphatic glands are certain to be affected if the disease has existed for any length of time.

6. The tumours are associated with the 'cancerous cachexia.'

Secondary Cancer.

The appearance of cancerous infiltration or tumours in parts distant from the original seat of the disease, is especially characteristic of the true malignant affection. Sometimes the secondary deposits do not occur until after the removal of the primary tumour; in other cases cancerous infiltration takes place independently of surgical interference.

Recent instances have occurred to illustrate this in a remarkable degree: secondary cancerous deposit effecting the entire pulmonary tissue having been traced to primary encephaloid in the spleen. In the same way epithelial growths upon the lip degenerating into encephaloid, may be followed by the development of disease in the lymphatic glands; although no operation has been performed upon the original tumour. The channels through which the morbid matter is conveyed to distant parts are problematical, but it seems that without the assistance of veins or lymphatics the process of contamination may proceed. The mere contact of a cancerous growth upon one organ with the surface of another organ, is sufficient in some instances to propagate the disease. Thus we sometimes meet with encephaloid of the liver causing encephaloid in that portion of the mesentery with which the organ is in intimate contact, thereby showing that the transudation of the morbid fluids into a healthy structure when the constitution is predisposed may establish an action which shall result in the production of new malignant growths.

PARASITES.

A PARASITE may be defined as a living organism growing upon or in the economy of another living organism.

There are both animal and vegetable parasites. The domestic flea may be taken as a rather unpleasant example of the former; while the mistletoe is an agreeable souvenir of the latter. An animal is here seen to live upon an animal, and a plant to be parasitic to a plant; but this is by no means necessarily the case, for an animal parasite may prey on the tissues of a vegetable, and a vegetable parasite have its habitat in the structures of an animal. Even vegetable and animal parasites are themselves in their turn liable to be the 'hosts'* of other parasites. Both kinds of parasites may coexist in the same organism.

Man, the boasting lord of creation, is preyed upon by a great variety of parasites, both animal and vegetable. To give the histology of parasites would be quite beyond the scope of these demonstrations, as the exhaustive pursuit of any one division of the subject is in itself sufficient to occupy a lifetime. Our object will therefore be merely to give a brief outline of the microscopic structure, habits, and mode of development of some of these parasites which are common to man, and the domestic animals.

Animal parasites are divided into the two great classes of
1. EPIZOA—Those dwelling on the exterior of their 'host,' and
2. ENTOZOA—Those infesting the internal organs.

The structure of the parasite is so markedly different in the two classes that a cursory examination will in most instances at once enable the observer to determine to which of the above classes the specimen belongs.

The detection of animal parasites is generally easy, most of them being visible to the naked eye.

Epizoa and their ova are found attached to hair, or among the desquamated scales from the cuticle, or in special canals or burrows. Many specimens require no preparation beyond being

* The term 'host' is used to denote the infested organism or habitat of the parasite.

transferred to the glass slide or animalcule cage, and subjected to moderate pressure. If not rendered by this method sufficiently transparent, they may be immersed in oil of turpentine, glycerine, or Canada balsam, made thin by the addition of turpentine. The antennæ, mandibles or other parts of these parasites may be dissected off and specially examined under the higher powers.

Entozoa are to be investigated by minute dissections under low magnifying powers at first, the amplification being increased as the more delicate parts are exposed.

Imbibition with carmine solution* is very important in the study of the internal structure of some entozoa. In flukes and in the segments of the tape-worm for example the coils and ramifications of the organs of reproduction can be very distinctly seen after imbibition, particularly when the preparation has been dried and mounted in thin Canada balsam. Acetic acid is also useful in rendering the exterior of the body of some animals transparent. Pressure between plates of glass may likewise be occasionally practised with advantage.

The ova, with their contained yelk in various stages of progress from commencing segmentation to the formation of the perfect embryo, may be obtained either from the fluids of the part which the entozoon infests, or by gentle pressure applied to the body of the parasite in the direction of the genital pore. They are usually best examined in water without any previous preparation.

The ramifications of the digestive tube are often apparent by virtue of their dark-coloured contents; their minute investigation, however, requires the aid of fine injections or imbibition.

The so-called water vascular system is also best seen in carefully injected or imbibed preparations.

Such being the general mode of procedure in the examination of animal parasites, we may now at once enter upon the special consideration of the different kinds.

EPIZOA.

ALTHOUGH the majority of epizoa, such as ticks, lice, fleas, and bugs, limit their depredations to the cutaneous surface, merely

---

* For the mode of using this solution see p. 22.

injuring it sufficiently to enable them to obtain the supply of blood requisite for their existence, there are many other kinds which not only penetrate the hair follicles and sweat glands, but even pierce and burrow beneath the integuments. Such, for example, is the case with the various forms of acari. Besides these, there are certain animals which are only epizoa while in the larval state. Thus, for example, the ova of the common house-, bluebottle-, and gad-fly, are deposited beneath the skin,* where they undergo development. The residence of the larvæ in this situation is only temporary, for as soon as the period arrives for their further development they are transferred elsewhere.

Those parasites which inhabit the organism temporarily while undergoing certain metamorphoses are sometimes included in a single class—Ectozoa.

Acari.

Acarus folliculorum, a small parasite found in the sebaceous

Fig. 191.

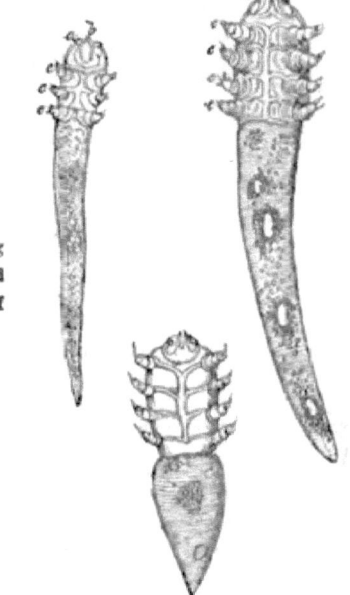

Acari from the cutaneous follicles. One young six-legged parasite, and two mature eight-legged specimens are represented. *a*. Papillæ on side of mouth. *b*. Mouth. *c*. Feet with bristles.

---

* The ova may be deposited in open sores as well as under the mucous membrane of the nose and frontal sinuses, where they often cause considerable irritation.

glands and hair follicles, from which they may be removed by squeezing the skin (of the nose and other parts of the face or trunk) between the thumb and forefinger. The exuded matter should then be placed on a slide with a drop of oil, gently spread out, covered, and examined under a high power.

The foregoing illustration (fig. 191) represents the three ordinary forms or degrees of development in which the acarus is observed in the human subject.

The presence of this parasite is commonly indicated by the occurrence of inflammation in the shape of minute pimples, or pustules.

Acarus scabiei, or itch insect, is perfectly familiar as the cause of a troublesome skin eruption associated with intense itching.

The mite takes up its residence at the end of a channel or burrow, which it forms in the skin at the side of the pustule. From this situation, which can be readily discovered by means of an ordinary hand lens, the insect can be removed by means of a fine needle; the operation, however, requires some little practice.

The mange-mite of the lower animals which produces the disease called scab in the sheep, and mange in the horse, dog, and cat, is closely allied to the itch-mite. It can be more readily detected, for, although it burrows beneath the surface, many specimens are constantly to be found upon the hair, and among the loosened epithelial scales of the diseased skin. A quantity of this débris is to be placed, in water, on a slide, under a low power, and the mites sought for. In most cases it will be necessary to disperse the matter over the slide by the aid of the dissecting needles, as the acari hide themselves under the scales, and although large enough to be visible to the naked eye when free from surrounding scabs and masses of epithelium, they easily escape observation when mixed with large amounts of foreign matter.

The female *Acarus scabiei* is distinguished by its larger size and the absence of any transverse band between the fourth pair of feet, which have no sucking discs. The presence of ova in the uterus will also be frequently observed.

The eggs are deposited in the burrows, and undergo development there. The young attain sexual maturity after changing their skins three times.

TICKS. 225

The male acarus is smaller than the female, and has the fourth pair of feet united by a transverse band, and furnished with sucking cups. It bores for itself small galleries or holes in the skin much smaller than those excavated by the female for the deposition of her eggs.

Of the two accompanying illustrations of acari, one is from the human subject, the true *Acarus scabiei*; and the other is from a sheep affected with scab.

Fig. 192.

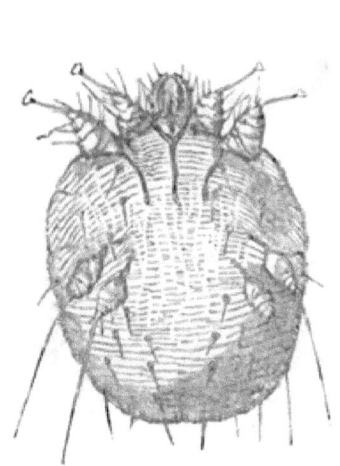

Acarus ovis.—$\frac{1}{50}$.

Acarus scabiei.

Ixoda.

Ticks are found commonly upon the skin of the dog and other domesticated animals.

The dog tick (*Ixodes Ricinus*) is distinguished by its somewhat oval body, with eight legs having terminal claws. The anterior part of the body is provided with a serrated apparatus, by means of which the parasite retains a tenacious hold of the skin.

The sheep tick (*Melophagus ovinus*) differs from the above in several particulars, belonging to another family; its habits are similar, but its mode of development is very singular, the embryos being formed and converted into pupæ before quitting the body of the mother.

The two illustrations which follow suffice to show the principal differences in form of the ticks infesting dogs and sheep.

Fig. 193.

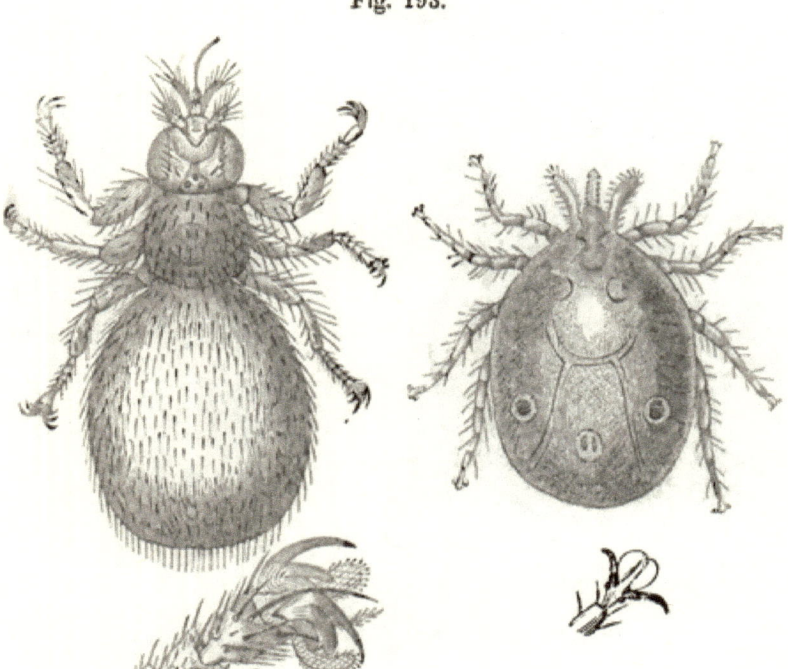

Ticks from the sheep and dog. The left-hand figure represents the parasite which is found on the skin of the sheep. One of the peculiar claw-like appendages belonging to the extremities of each parasite is figured below (magnified 100 diameters).

During the process of feeding, the tick buries its sucking apparatus in the animal's skin and fills itself with blood until its abdomen becomes distended to several times its natural size. While thus fixed the parasite may be torn asunder without loosening its hold.

Pediculi.

The common louse infesting the heads and bodies of the human subject, and the varieties of it common in sheep, dogs and horses, belong to this family. They are distinguished by six legs; three on each side the thorax, and two antennæ. The abdomen is divided into segments.

These parasites are easily detected from their size; their ova are also discernible clinging to the hairs.

The crab-louse lives among the hair of the pubis, chest, axilla, and even in the eyebrows, but never in the head; it bites deeply and produces considerable irritation. It is distinguished from the common louse by its fiddle-shaped head, and broad flat thorax. The ova of both are to be found adhering to the hairs.

The drawings of the three parasites will suffice to render the differences of form apparent.

Fig. 194.

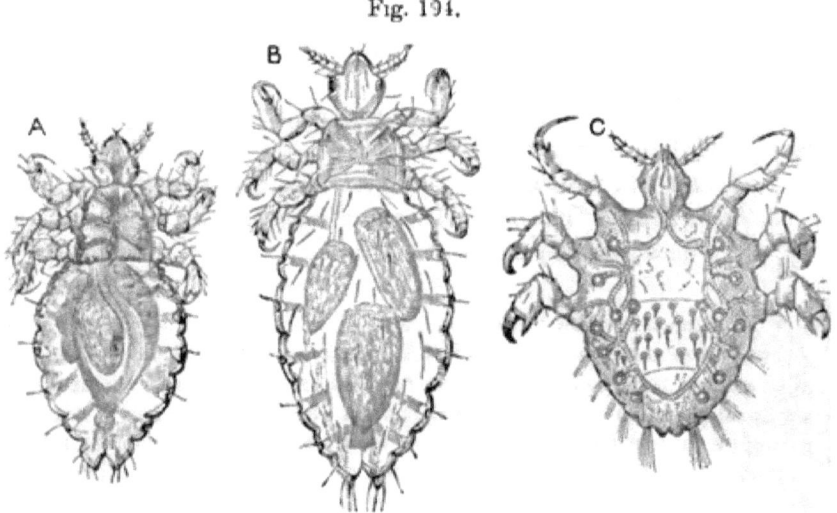

Pediculi from head, body, and pubis of human subject. A. Head-louse. B. Body-louse. C. Crab-louse.

ENTOZOA.

NOTWITHSTANDING considerable diversity in the forms, structures, and methods of development of the various internal parasites, it is convenient to include them under the general term of Entozoa, for the simple reason that they all agree in their habitat, being the interior of the organism of another animal. Some internal parasites multiply within the host whose tissues they infest; others never advance beyond the larval condition until they change their habitation. They then attain to sexual maturity, and become capable of propagating their kind by the development of eggs containing embryos, which, being expelled, are subsequently conveyed by various means into another organism, where they assume the larval form. After a time these larvæ in their turn change their host and become mature animals. This process of change is characterised by the term 'alternate generation.'

Entozoa have been divided into three great orders, or classes:*—

1. CESTODA, which includes all the various forms of tape-worms in their different stages of development;

2. NEMATODA, which includes all the round worms;

3. TREMATODA, fluke-shaped worms, including all that possess openings on different parts of their bodies.

* To these orders have been recently added the TURBELLARIÆ and the ACANTHOCEPHALÆ; the Turbellariæ being distinguished by the possession of ciliæ all over their bodies. The order comprises—1. *Planariæ*, which have a general resemblance to flukes, and some of which have a power of propagating by natural fission; 2. *Nemertes*, or ribbon-worms, resembling tape-worms, but having an extraordinary power of extension; the usual length is five or six feet, but a specimen is described twenty-two feet long, and the animal is said to be capable of extending itself to eight times the length it occupies when at rest.—Acanthocephalæ includes all worms whose anterior portion is furnished with a proboscis surrounded with hooks disposed in circles. These parasites are found in all kinds of animals save man; in birds and fowls they are abundant. The intestines of the lesser water-newt commonly contain numerous specimens of the *Echinorhynchus anthuris*, which will serve to illustrate the peculiarities of this order.

Cestoda—Tape-Worms.

Tape-worms are distinguished by their long, soft bodies, being divided into joints, or segments. The head is, as a rule, furnished with four suckers, and generally has also a peculiar arrangement of hooks. The posterior joints, or mature segments, have a distinct sexual system belonging to each, and enjoy for some time an independent existence after being separated from the worm and expelled from the infested animal. The eggs which they subsequently void are introduced in a variety of ways into the bodies of other hosts, where the little embryo with its six hooklets bores its way through the tissues until it reaches a locality well adapted for its habitat. Here it assumes the larval form, better known as that of the hydatid, but beyond this stage of development it never advances in this locality. Before it can become a perfect tape-worm it must be again transplanted to the intestinal canal of some other animal. As soon as this occurs, the head of the hydatid attaches itself to the intestinal mucous membrane, and while its caudal vesicles gradually shrivel up and drop off, true tape-worm segments begin to appear. These grow and multiply with such rapidity that the perfect animal is soon reproduced; and so soon as its posterior segments become mature they are, in their turn, detached and expelled with the fæces. Thus, commencing with the ovum, we have three well-marked phases to distinguish in the development of cestoid parasites.

There are several varieties of tape-worms, each being distinguished by structural peculiarities relating to the form and disposition of the hooks, the size and shape of the segments, and the arrangement of the sexual organs.

In describing these worms authors commonly make use of the terms Strobila, to signify the entire worm, sexually mature; Proglottis, the mature segment or joint; Scolex, the hydatid or larval form, including the tape-worm head; and Proscolex, the embryo contained in the ovum.

Tænia solium.

Strobila or Mature Worm	Scolex or Hydatid	Proscolex
Tænia solium. Common tape-worm of man.	Cysticercus cellulosæ. Measles of the pig.	The embryo contained within the ovum, and furnished with a boring apparatus of hooklets, six in number, arranged in pairs. Found in human excrement, and on field and garden stuffs.

Fig. 195.

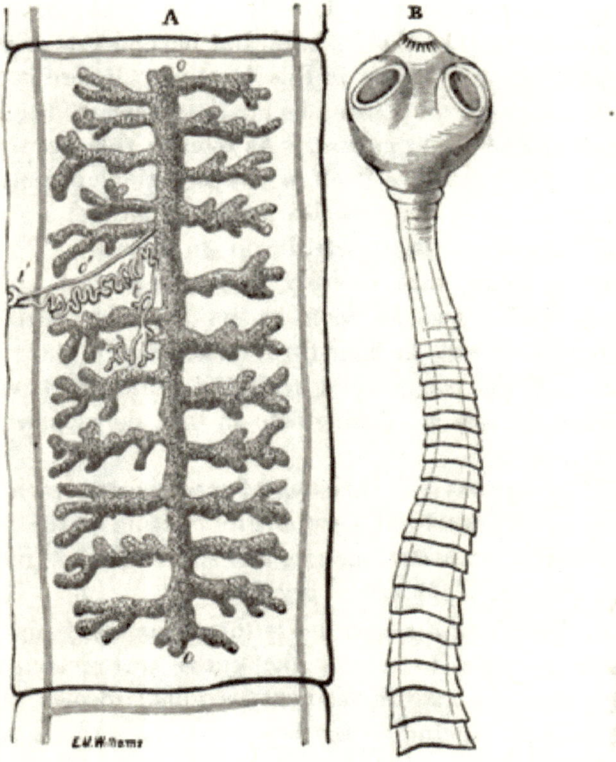

Tænia solium (from Blanchard). A. One of the longer mature posterior segments with the sexual organs fully developed. *o, o.* Ramified ovary full of ova. *o'.* The oviduct. *t.* The tubular testis. *t'.* The penis, &c. B. Head, neck, and anterior recently formed segments.

Tænia solium infests the intestines of man. Single specimens are often found, but two or three are not uncommon, and considerable numbers now and then exist. The worm varies in length, from ten to twenty feet or more. The head is very minute, and has four suckers, above which occurs a circular projection (rostellum) surrounded by a double row of hooks. The posterior mature segments contain a branched uterus; the genital openings are laterally placed in an irregularly alternate manner. These several parts are indicated in the drawings (fig. 195).

The hooks, examined separately under a high power, differ in shape in the various species of tape-worms sufficient to afford one aid to their distinction from each other, and to connect them with their larval or hydatid forms, which occur in situations altogether different from those selected by the mature parasites. Thus the relation between *Tænia solium* of the human subject and the *Cysticercus cellulosæ* (or measles) in the flesh of the pig, was first deduced from the discovery of the identity in form of the hooks of the hydatid and the tape-worm; and in every other instance the comparison holds good. The facts, however, are now placed beyond dispute by numerous and carefully conducted experiments. Animals fed with tape-worm segments containing mature ova, are found to become the subjects of hydatids, which, being given to other animals, advance to the condition of tape-worms. These experiments, to be successful, must be conducted with due regard to the species of tape-worm or hydatid with which each animal is infested naturally, otherwise no positive results will follow.

Cysticercus cellulosæ is found sometimes abundantly in the muscles of the pig.* Each hydatid, when fully developed, is about the size of a bean, and consists of a flask-shaped vesicle (caudal vesicle) filled with albuminous fluid, and having a dense white spot at one point; this is the receptacle in which the head is lodged, and from which it may be extricated by means of fine needles. In the dead hydatid it is commonly found protruded. Under a low power the suckers, rostellum, and double crown of hooks will be readily seen, and their exact resemblance to the same structure of the *Tænia solium* at once established.

---

* Not long since a specimen was found attached to the testicle removed from a pig by castration.

· The *Cysticercus cellulosæ* occurs in man in the brain, eye, and other parts from the accidental swallowing of mature eggs of *Tænia solium*; in this way an individual harbouring tape-worm may infect himself with hydatids, by inadvertently swallowing the eggs from expelled mature segments.

Cysticercus cellulosæ is also said to be found in the dog, bear, and other animals.

Tænia mediocanellata.

Strobila	Scolex	Proscolex
Tænia mediocanellata. Met with in man.	Cysticercus of measles, affecting the muscles of calves and oxen.	Embryo contained in mature ovum. Found on the herbage on which the cattle feed.

One of the segments of the Tænia and the head of the worm are represented in the woodcut.

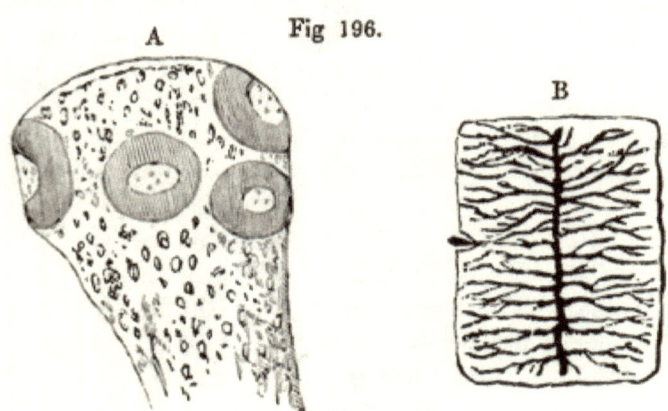

Fig 196.

A. Head of Tænia mediocanellata, with the four sucking discs coloured with black pigment. B. One of the mature proglottides.

Tænia mediocanellata appears to be almost as common as the *Tænia solium*, with which it is often confounded.

The worm is distinguished by its greater size; the arrangement of the uterus, which gives off a larger number of lateral branches than are found in the *Tænia solium*; the existence of the genital pores in a lateral position, below the central line of each mature segment; and still more definitely by the conformation and aspect of the head, which possesses neither hooks nor rostellum. The four sucking discs contain an abundance of pigment granules, rendering them nearly black, and giving to the head a very characteristic appearance.

The hydatid cysts of *Tænia mediocanellata* are developed abundantly in the muscular structure of calves fed on the mature segments. The Cysticerci attain to a somewhat larger size than those of the pig, and possess the peculiar heads of *Tænia mediocanellata*.

Tænia marginata.

Strobila	Scolex	Proscolex
Tænia marginata (Tænia tenuicollis). Habitat, dog and wolf.	Cysticercus tenuicollis. Of man, cattle, sheep, and other animals.	Embryo contained in mature ovum. On garden and field produce.

Tænia marginata does not occur in man, but in the dog. It is very common from the fact of its scolex being almost constant in the omentum and other internal parts of the sheep, frequently hanging loose into the abdominal cavity; being of considerable size, it is cut away by the butcher and cast out without any precaution. The custom of feeding dogs upon the internal parts of sheep without previous cooking frequently occasions the introduction of the scolex.

The *Cysticercus tenuicollis* seems to produce little inconvenience in the animal which it infests. In one remarkable instance, a fine healthy lamb, three months old, was slaughtered by the butcher, and was found to contain countless numbers of cysts attached to every portion of the abdominal cavity and its viscera; several dogs were fed upon the infested organs, and in every case the intestines became filled with *Tænia marginata*.

Tænia echinococcus.

Strobila	Scolex	Proscolex
Tænia echinococcus. Dog and wolf.	Echinococcus (often called pill-box hydatid, acephalocysts, &c.). Is common to man, horses, cattle, sheep, and various animals.	Embryo contained in the mature ovum. Garden and field produce.

Tænia echinococcus is distinguished by its extreme minuteness: it consists of a head with the usual sucking discs, hooks, and rostellum, and three segments, the last but one being the longest and containing the sexual organs.

The woodcut will convey some idea of the form and development of the scolices within the parent vesicle.

Fig. 197.

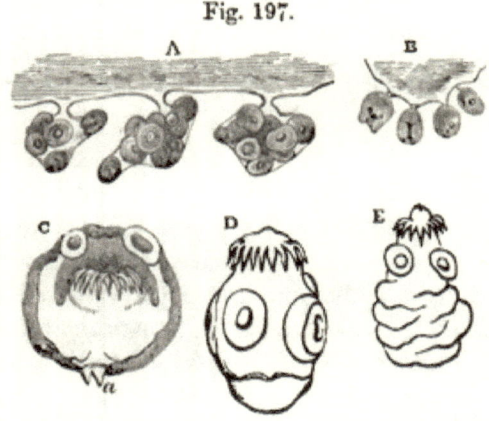

Echinococcus hominis (from Wilson). A, B. Grouped and single Echinococci, attached by peduncles to the inner membrane of the cyst. C. A contracted, and D, an expanded Echinococcus. *a.* The peduncle. E. A more advanced animal, shrivelled.

Echinococcus cysts, which are developed from the six-hooked embryo of the mature Tænia, occur in various organs, the liver, lungs, kidneys, bladder, bones, neck, and other parts.

Echinococci are distinguished from all other hydatids by the existence, in the interior of the cysts, of myriads of minute scolices or tape-worm heads, which are either found free and floating in the fluid of the cyst, or included in brood capsules attached to its walls, or otherwise within secondary vesicles (daughter and granddaughter vesicles), formed by a process of budding from the parent cyst.

In a recent case, in University College Hospital, the fluid taken from an apparently serous cyst in a woman's neck was found, upon microscopic examination, to contain a number of Ecchinococci, thus proving the tumour to be a hydatid instead of an ordinary serous sac. The cysts are sometimes voided entire along with the urine.* Ecchinococcus cysts are often found in the liver of the horse and ox, sometimes in such numbers that scarcely any of the gland structure remains.

Tænia serrata.

Strobila	Scolex	Proscolex
Tænia serrata. Dog.	Cysticercus pisiformis. Rabbit.	Embryo in the mature ovum. Garden and field produce.

Tænia serrata is found in the intestines of the dog, sometimes of the length of four feet. Its segments are narrow in proportion to their length, and the margin produced by their junction has a serrated character.

The scolex of this worm, *Cysticercus pisiformis*, is found constantly in the livers of rabbits; and frequently hanging loose into the abdominal cavity. The cysts are small in size and often occur in great numbers.

Tænia crassicollis.

Strobila	Scolex	Proscolex
Tænia crassicollis. Cat.	Cysticercus fascio- laris. Mouse.	Embryo in the mature ovum.

* *Medical Times and Gazette*, March 25, 1865.

Tænia crassicollis, found in the intestines of the cat, does not exceed five to seven inches in length; the head is usually large, the suckers often being visible to the naked eye. The worm varies but little in breadth from the head to the termination of the segments.

Cysticercus fasciolaris, the larva of the Tænia, exists in the liver of the mouse. The cyst is sometimes so small as to be scarcely perceptible, sometimes it is found the size of a pea; but its special peculiarity is seen in the assumption of the form of the tape-worm while it still remains a hydatid cyst, as shown in the drawing, which also represents the head and some of the segments of the mature Tænia.

Fig. 198.

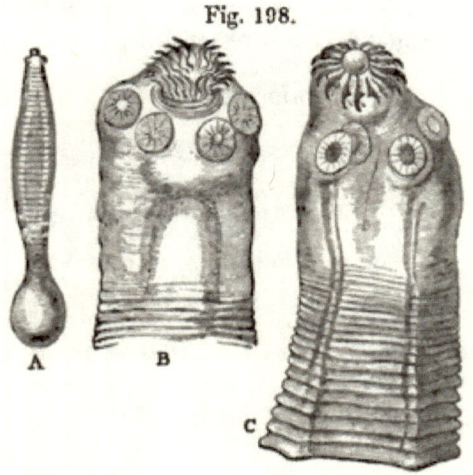

Cysticercus fasciolaris of the mouse, and Tænia crassicollis of the cat. A. Cysticercus fasciolaris from the liver of the mouse, twice the natural size. B. The head of the same, magnified (from Dujardin). C. Head and first segments of the body of Tænia crassicollis of the cat, showing the double circle of hooks, a few of the smaller under circle being seen where one or two of the larger ones have fallen off.

Tænia cænurus.

Strobila	Scolex	Proscolex
Tænia cænurus.	Cænurus cerebralis.	Embryo in the mature ovum.
Dog.	Cattle, sheep, horse, goat, and rabbit.	Field produce.

Tænia cœnurus of the dog is the mature form of a peculiar hydatid, very common in the brain of sheep, producing the disease known as 'staggers' or 'gid.' The hydatid, instead of having one head, possesses several hundreds, attached to the caudal vesicle. Generally the little eminences in which the heads are contained are found projecting into the interior of the vesicle; they may, however, be pressed out, or teazed out with needles, or, if the vesicle is left in weak spirit for some time, they will be everted naturally. Under the low power the suckers, rostellum, and hooks, will be easily perceived.

Tænia elliptica—Tænia cucumerina.

Strobila	Scolex
Tænia elliptica, or cucumerina. Man and cat and dog.	Unknown.

Tænia elliptica of the cat, and *Tænia cucumerina* of the dog, are probably identical. The parasite reaches the length of six or seven inches, and is distinguished by its narrow oval or elliptical segments, each one having two genital pores, one on each margin directly opposite to each other. The hooks of this worm very readily fall off, but perfectly fresh specimens will be found to possess the usual double row round the rostellum.

A large number of these worms fully grown were recently expelled from a puppy six weeks old after a dose of areca nut.

Bothriocephalus latus.

Strobila	Scolex
Found in man, sheep, dog, and cat.	Unknown.

This parasite differs from other tape-worms in some important particulars. The head is unfurnished with hooks, and the four suckers are replaced by two lateral fissures. The worm attains, sometimes, the length of twenty-five feet, and, at the broadest parts, nearly an inch in width. The genital pores are found in the central line instead of on the margins of each segment, and frequently there are two openings on the same joint, close together.

In one instance, the intestines of a sheep were found to be filled with these parasites, and nearly all of them had the two genital pores very distinctly marked.

Although the Bothriocephalus is more common in Switzerland than elsewhere, we occasionally meet with it in natives of this country. It is also now and then to be met with in the cat and dog, but in these animals the specimens, although perfectly mature, are usually of small size.

The accompanying illustrations represent the head and the central genital pores of this parasite.

Fig. 199.

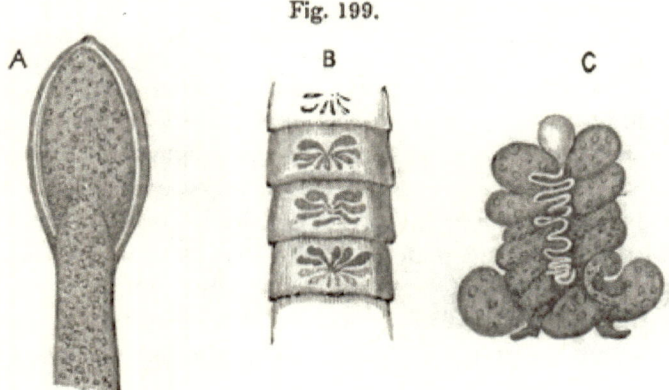

Bothriocephalus latus. A. Head, with lateral apertures. B. Segments with the uterine coils in the centre. C. Generative organs.—Magnified 200 diameters.

NEMATODA—ROUND WORMS.

All varieties of round worms are included under this head, and, accordingly, the class is a large one. With few exceptions the parasites are of small size, and commonly exist in enormous numbers. Comparatively little is known of their developmental changes or the mode of their introduction into the animal organism.

Ascaris lumbricoides inhabits the human intestines, sometimes in considerable numbers. It occasionally migrates to the stomach, and sometimes even passes up the œsophagus, and makes its appearance in the mouth and nostrils.

The worm varies in length from six to fourteen inches; its body is finely ringed, and the head is furnished with a mouth having three papillæ.

The male is distinguished by its smaller size and double penis. In the female the coils of the ovarium, which is very long and tortuous, may be seen through the transparent parietes. The genital opening is situated near the middle of the worm.

Two other varieties of Ascaris are found in the human intestines, *Ascaris mystax*, commonly found in the cat, and *Ascaris vermicularis*, better known as *Oxyuris*.

The *Ascaris megalocephala*, which closely resembles *Ascaris lumbricoides*, is found abundantly in the intestines of the horse. It is larger than the human species.

The presence of this parasite in the intestines of the horse does not seem to produce any great inconvenience, and were it not that occasionally some are voided no suspicion would in many cases be entertained that the animal was infested; and very frequently when horses have been examined after death the intestines have been found to contain immense numbers of these parasites, the presence of which was not suspected during the animal's life. In the illustration it will be observed that some of the uterine coils have protruded through the integument; this is very frequently seen in all the members of this family.

For the purpose of examining these large worms microscopically, it is necessary to dissect them under a low power, and to prepare the different parts separately for the higher powers.

240 MORBID HISTOLOGY.

The drawing represents a specimen of *Ascaris* from the intestines of a horse.

Fig. 200.

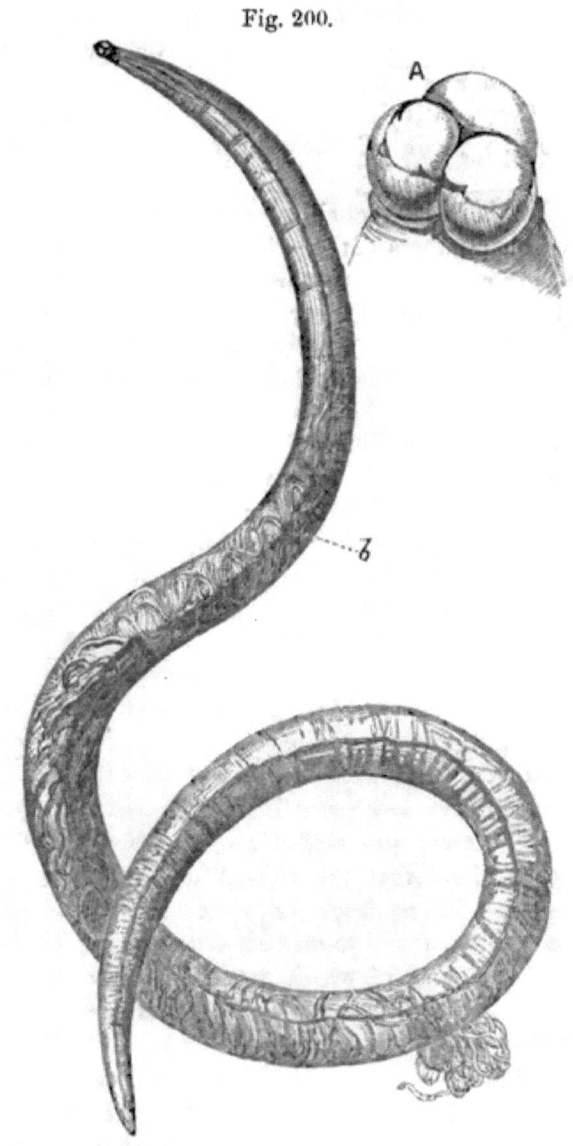

Ascaris megalocephala (horse). A. Mouth, with the three papillæ, magnified 100 diameters. *b.* Coils of ovarium, seen through the integument.

ROUND WORMS.

Oxyuris vermicularis.—A small thread-worm infesting the rectum, especially in children, causing much annoyance by its nocturnal wanderings. Sometimes found in the nostrils.

The male is less than a quarter of an inch in length; the female about double that size.

Oxyuris is characterised by a body of fusiform shape, terminating in a tapering tail.

The drawing represents *Oxyuris vermicularis*.

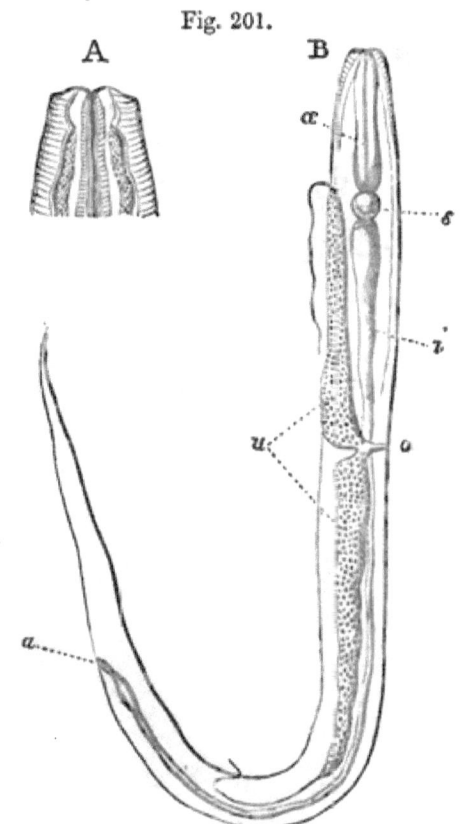

Fig. 201.

Oxyuris vermicularis. A. Head of the parasite, more highly magnified. B. The entire worm, magnified 100 diameters. *œ*. Œsophagus. *s*. Stomach. *i*. Intestine. *o*. Oviduct. *u*. Uterus. *a*. Anus.

Filaria bronchialis, or **Strongylus bronchialis**, is found in the bronchial tubes of man, and, when accumulated in vast numbers, is capable of causing suffocation.

In the lungs of the sheep a variety of this worm is frequently found occupying little tuberculous-looking spots, commonly mistaken for tubercle (from which the lungs of the sheep appear to be perfectly exempt). The young of the parasite migrate to the small bronchial tubes, and occasion considerable irritation.

The drawing represents one of the so-called tubercular masses from the lung of a sheep, and shows the worm coiled up amidst a quantity of exudation matter, undergoing calcareous degeneration.

Fig. 202.

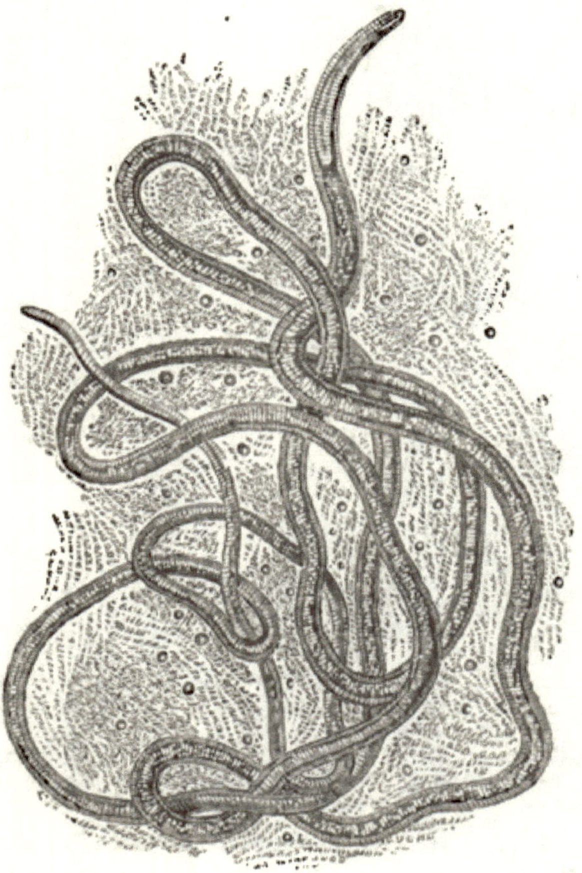

Strongylus Filaria from the lungs of the sheep.

In calves *Filariæ* are very common, producing the disease known as 'husk,' from the constant cough which is present.

Filaria oculi.—Minute specimens of these worms have occasionally been met with in the crystalline lens and the anterior chamber of the eye, floating in the aqueous humour. The parasites appear to occur most frequently in the eye of the horse, although now and then they are met with in the human subject.

The structure and development of the worm have not been satisfactorily made out, but some of the best authorities consider it to be the sexually immature young of some *Filaria* that has migrated to this position.

Filaria Medinensis, or **Guinea Worm**, occurs only in the tropics. It is found most frequently beneath the skin in the cellular tissue of the legs, although in the natives of Hindostan, who wash after defæcation, it is occasionally found in the neighbourhood of the genital organs, and also between the shoulders of water-carriers, who transport their burdens in skins on their backs.

Nearly the whole of the worm is occupied by the uterus, which is generally filled with the young ones, which are distinguished by their thin tapering tails, as shown in the drawing.

Fig. 203.

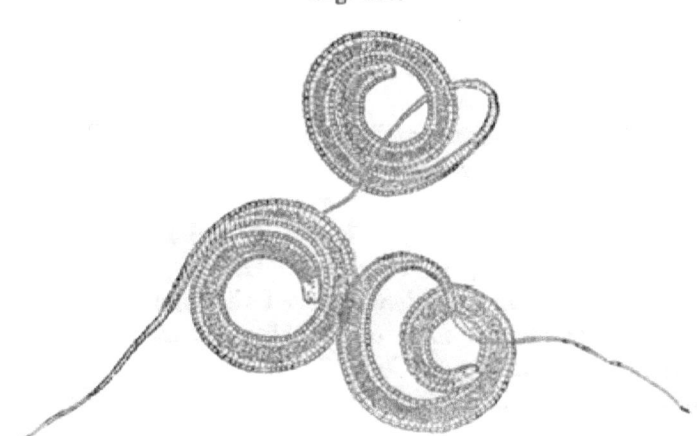

Young of the Guinea worm.—Magnified 200 diameters.

The female worm only is known, and reaches an average length of three feet, although it is stated to vary from one foot to ten or twelve feet. The body is cylindrical, tapering at both ends, and about one-eighth of an inch in thickness.

The young of the Guinea worm inhabit pools and marshy places, and find entrance into the human body, not by the mouth, but by boring through the skin, or entering some of the follicles. Once located there, they grow with rapidity, attaining their full length in from five to ten months, about six being the usual time.

Strongylus gigas is a large species of round worm infesting the kidneys of man and some of the lower animals. The female grows to the length of two or three feet, the male to a foot long; the colour of the worm is reddish. Of its mode of reproduction nothing certain is known.

Strongylus paradoxus is found in the bronchial tubes of the pig. The female worm is about one inch and a half long, and is peculiar from the constant protrusion of numerous coils of the reproductive organs, filled with eggs and embryos, many of them living, from different parts of the body; sometimes the aperture is near the head, and sometimes close to the tail, and occasionally there are two or three openings in the same worm.

The male worm is smaller than the female, and is known by the bursal appendage at its tail and its long double hair-like penis.

Trichocephalus dispar, well known by the name of the whip-worm, from its peculiar form, is found principally in the human cæcum. The worm varies in length from an inch and a half to two inches; the neck is double the length of the body, and much less in diameter, being, indeed, a mere filament.

For some time the opinion obtained that the *Trichocephalus dispar* was the mature form of *Trichina spiralis*; this is now known to be erroneous.

The illustration shows the *Trichocephalus* of its natural size and magnified.

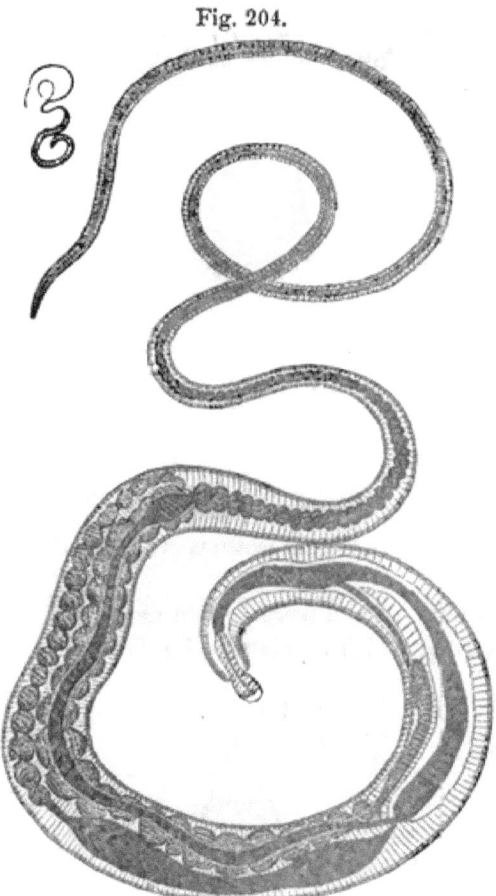

Trichocephalus dispar.

Trichina spiralis.—Much interest attaches itself to this parasite, in consequence of the extraordinary prevalence of the young brood in the muscles of the pig, from which, unfortunately, it is easily transferred to the intestines of man. After this migration the worm rapidly attains sexual maturity, and develops a number of embryos, which immediately wander from the intestines to the muscles, so that in a very brief period almost every voluntary muscle in the body becomes infested.

In their larval condition *Trichinæ* are found coiled in a spiral form in small cysts in the muscle, presenting to the naked eye the appearance of minute specks, as shown in the woodcut.

Fig. 205.

Cysts of the Trichina spiralis in situ.—Natural size.

The appearance of the worm, when magnified, is indicated in the next drawing, which represents the *Trichina* within its cyst.

Fig. 206.

A separate cyst of the Trichina, which is seen coiled up through the transparent coats.—Magnified 200 diameters.

After introduction to the intestines of man or some other warm-blooded animal, the young brood becomes mature in two days, and in four more days the females contain embryos ready to pass out and migrate to the muscles of the animals they infest, a remarkable example of the occurrence of the larval forms and mature parasites in the same host. In about two months after they get into the muscles, they become encysted. In two months more the cyst calcifies, and in two more the worm itself suffers the same fate.

A drawing of the sexually mature *Trichina* is given below.

Fig. 207.

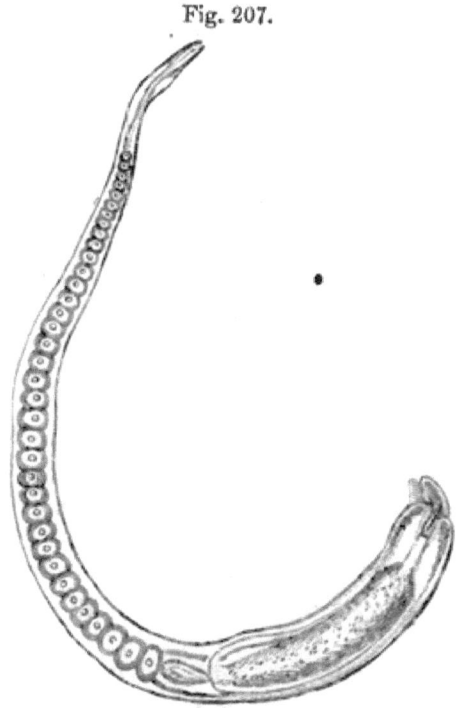

Mature Trichina, found in the intestines.

The *Trichina* is much more common in this country than is generally supposed. It simply escapes notice in consequence of not being looked for.

Sclerostoma syngamus is a small worm infesting the tracheæ of birds, and causing the well-known disease 'gapes.' The female parasites are less than three-fourths of an inch long, the males not more than one-eighth.

The male is usually found with the lower end of the body attached to the vagina of the female; it is even asserted that this position becomes permanent by a positive growth of the two parasites together.

Sclerostoma duodenale, or **Anchylostomum duodenale**, is not unlike the variety inhabiting the tracheæ of birds, save that the two sexes are not found attached together.

The worm is common in Egypt and Northern Italy, and infests the small intestines of man, attaching itself firmly to the mucous membrane, and producing by its presence considerable systemic derangement.

The head of the worm is provided with a mouth having three horny papillæ, or teeth. The tail of the female is pointed; that of the male terminates in a membraneous bursa having eleven rays.

Pentastoma, so called from the five dark marks in the vicinity of the head having been considered to represent so many openings; the appearance, however, is in reality due to a pair of hooks on either side the mouth.

Pentastoma tænoides is found in the nostrils and frontal sinuses of the sheep, dog, and some other animals. *Pentastoma denticulatum*, which is the larval form, inhabits the abdomen, being commonly fixed to the surface of the lungs or the liver.

A very large variety of the parasite was found by Dr. Harley in an Egyptian cobra. Two of the worms had penetrated the air-sacs, and fixed themselves into the walls of one of the large bloodvessels, and were living apparently upon the animal's blood.

TREMATODA—FLUKE-SHAPED WORMS.

The common liver-fluke is an instance of this class of parasites, which includes a large number of soft-bodied entozoa having openings on different parts of their bodies.

Flukes are found but rarely in man. In the sheep they are found to be exceedingly abundant in the disease known as 'rot' or 'bane.'

Varieties of the parasite are found in birds, reptiles, and fishes.

Distoma hepaticum, or **Fasciola hepaticum**, the fluke infesting the liver of man and many of the lower animals, is aptly compared to a small sole in shape. In size, however, it does not exceed an inch and a quarter in length, and three-quarters of an inch in breadth, when fully grown. The majority of specimens are less than this. The surface of the body is armed with spines, which are easily distinguished by the aid of a common lens.

The drawing of the parasite will convey an idea of its general appearance, and the arrangement of its internal organs.

Fig. 208.

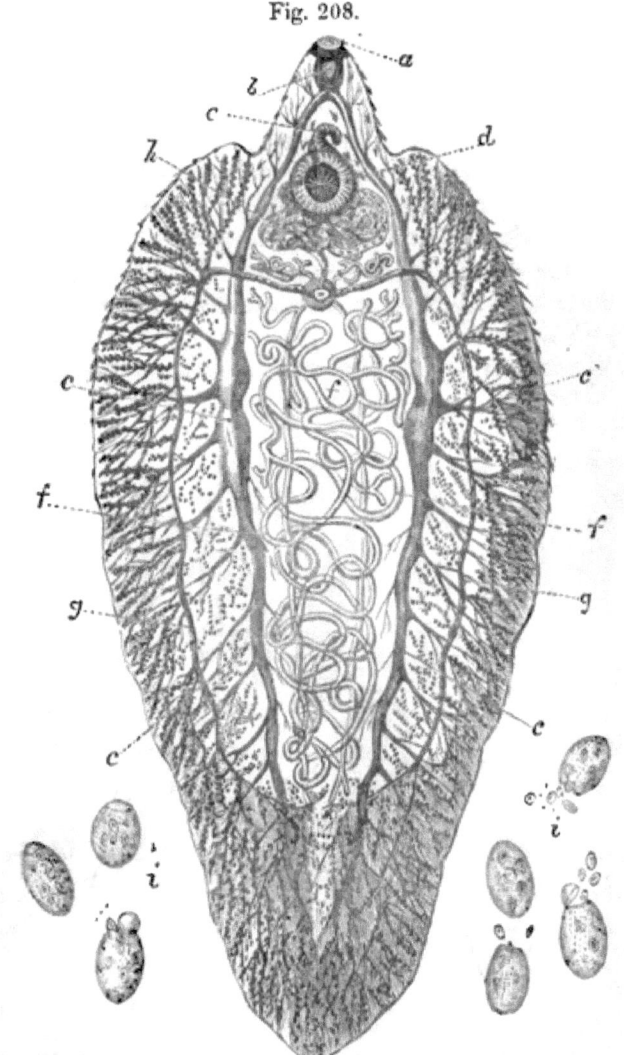

Liver Fluke. *a.* Anterior or oral opening, communicating with the œsophagus (*b*) and digestive tubes (*c*). *d.* Genital opening with the curved male organ (*e*) protruded. *f.* Male generative organs. *g.* Oviducts leading by transverse branches to the uterus (*h*). *i.* Various ova with their opercula and granular contents.

The mature eggs of the liver-fluke, it is quite certain, do not undergo further development until they are expelled from the host, when, falling upon marshy or moist places, the embryo shortly becomes free, and moves about with activity by virtue of its surrounding ciliæ.

The changes which subsequently take place in the course of the development of *Fasciola hepatica* have not yet been traced, but from observations made upon other varieties of Distoma, there is every reason to believe that the embryo succeeds in entering the body of some molluscous animal, in which it assumes a larval form, and develops other larvæ in its interior; these again, when free, produce still higher larval forms, which are known as *Cercaræ*; these last pass through a pupa state, and emerge from their cysts as flukes.

The illustration represents the several stages of development.

Fig. 209.

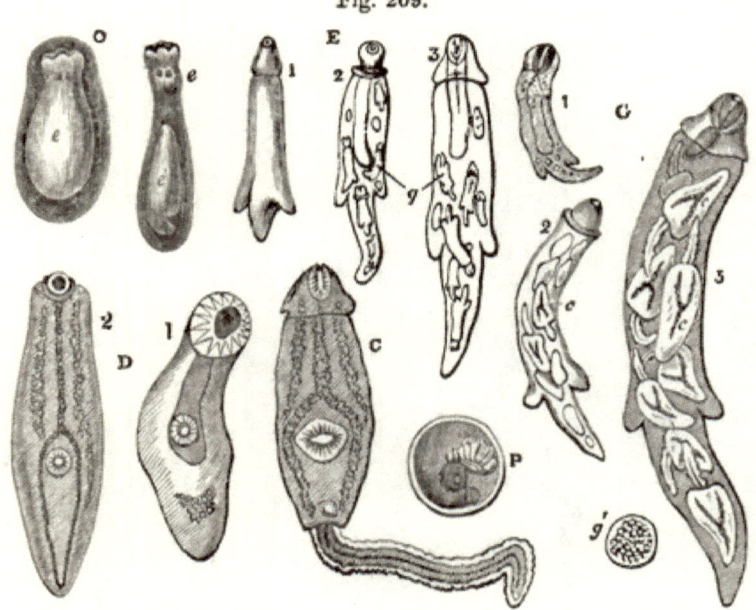

Series of changes in the development and generations of Distoma (from Steenstrup). O. Ovum with embryo. *e*. This embryo in a free moving state. *e'*. Another embryo in its interior. E. This last embryo farther advanced. 1. First stage, soon after it becomes free. 2 and 3. Farther on, with *g*, the second generation, within them in various stages. G. 1. One of this second generation at an early period. 2 and 3. Farther on, with *c*, *e*, Cercariæ, or Distoma larvæ, within them. *g'*. One of the granular globules from which the Distoma larvæ and previous generations arise. C. One of the Cercariæ, or Distoma larva. P. The same, passed into its encysted or pupa state. D. Distomata. 1. Young Distomata. 2. Distoma found deep in the viscera.

The manner of introduction of *Distoma* into the body of man and the lower animals is obviously by the accidental swallowing of small molluscs upon herbage and vegetables; such molluscs containing in their bodies some of the higher larval forms of *Distoma*, probably the *Cercariæ* in the pupa state. Once within the body of the host, the parasite finds its way to the liver, and occupies the gall-ducts and gall-bladder. The mature eggs are expelled by thousands, and may be discovered in the bile, and also in the ingesta with which that secretion is mixed.

Distoma lanceolatum.—A small lancet-shaped fluke, not quite half an inch in length, and less than a quarter in breadth. The arrangement of the digestive canal is much more simple than that of the liver-fluke last described, as, instead of branching out laterally, it merely divides into two tubes, which terminate in blind extremities.

Distoma lanceolatum has been discovered in the liver of the human subject in two or three instances only. In the sheep it is more common; but the liver of the ox is said to be its most usual habitation. It has also been seen in cats, deer, hares, and rabbits.

Distoma opthalmobium.—A minute parasite that has been seen in a few cases in the disease of cataract between the crystalline lens and its capsule. The largest specimen that has been found did not exceed $\frac{1}{20}$ of an inch in length. The parasite is considered to be the immature young of some species of fluke, and not a perfect distome.

Distoma hæmatobium is peculiar on account of the male and female sexual organs being contained in separate individuals. The male is vermiform, and about half an inch long; the female rather longer and thinner.

The parasite is common in Egypt, the East and West Indies, South America, the Mauritius, and the Cape of Good Hope, and infests the portal bloodvessels as well as the kidneys, ureters, and bladder, giving rise to the well-known endemic hæmaturia of the countries above mentioned. The detection of the peculiar ova of the worm in the urine at once serves to diagnose the nature of the malady.

The drawing represents the two sexes of *Distoma hæmatobium*, and also some of the eggs.

Fig. 210.

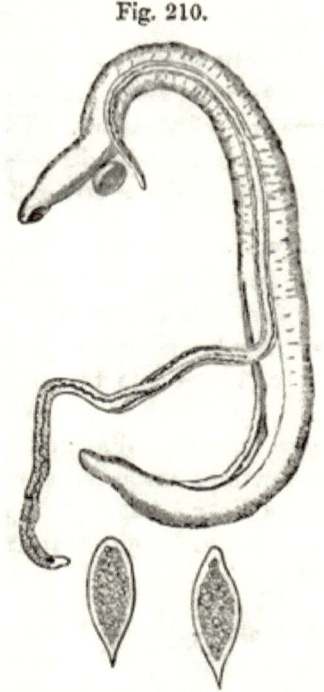

Distoma hæmatobium, male and female, also two ova.

Distoma heterophyes.—A minute distome of pyriform shape, about $\frac{1}{16}$ of an inch long, having a number of spines over the surface of its body. These parasites have been discovered in a few instances in the small intestines of the human subject.

VEGETABLE PARASITES.

Two great classes of vegetable parasites infest the animal organism—FUNGI and ALGÆ.

Those occurring on the surface of the body are named EPIPHYTES, and correspond to Epizoa. Those inhabiting the interior of the frame are called ENTOPHYTES, and correspond to Entozoa.

Although an immense number of species of fungi and algæ have been described, it now appears that the great majority of them are mere varieties of a very limited number of species; the variations they present depending upon the different conditions under which they accidentally grow.

Fungi and algæ are often confounded together in consequence of some of the number of each group consisting of filaments, with a reproductive system of spores enclosed in a 'sporangium,' or resting upon a 'receptaculum.' But the algæ are all more or less coloured, in consequence of their containing chlorophyll, or some other pigment. Single cells, or filaments of algæ, however, commonly appear colourless, and before attempting to distinguish them, it is necessary to view them when aggregated in masses.

Vegetable parasites generally require to be examined under a high power in order to make out their minute structure, although, when considerable masses are collected, as in the case of common mould, a good idea of the general arrangement is to be gained by the use of a low power.

A drop of caustic potash should be allowed to flow under the covering glass, to clear up any adventitious matter, and render the several parts of the fungus more distinct.

If spores only are found, it is desirable to keep them in water, or in the fluid in which they were discovered, in order that they may germinate, and enable the observer to determine the species.

Algæ and fungi are to be sought for upon the integument, among the hairs and crusts, as well as in discharges from wounds and sores; and also in vomited matters. The frothy surface of a fluid should always be submitted to a careful examination, as the parasite is more likely to be there than at the bottom.

FUNGI.

Oidium albicans.—The well-known thrush-fungus is developed on the mucous membrane of the mouth, tongue, and œsophagus, in aphtha, as well as on the surface of sores and diphtheritic exudations. It is a fungus found almost everywhere. It grows equally well in aqueous solution of strychnine or oxalic acid as in the most benign liquid. Its spores are constantly floating about in the atmosphere, and the fungus consequently springs up wherever it finds a suitable habitat; a moderate heat and moisture, in conjunction with the presence of animal or vegetable matter, being all that is necessary for its development.

Oidium albicans, as will be seen by reference to the illustration, consists of tubular filaments, of branching stems, and of numerous minute spores. Specimens taken from the mouth are generally mingled with masses of epithelium.

Fig. 211.

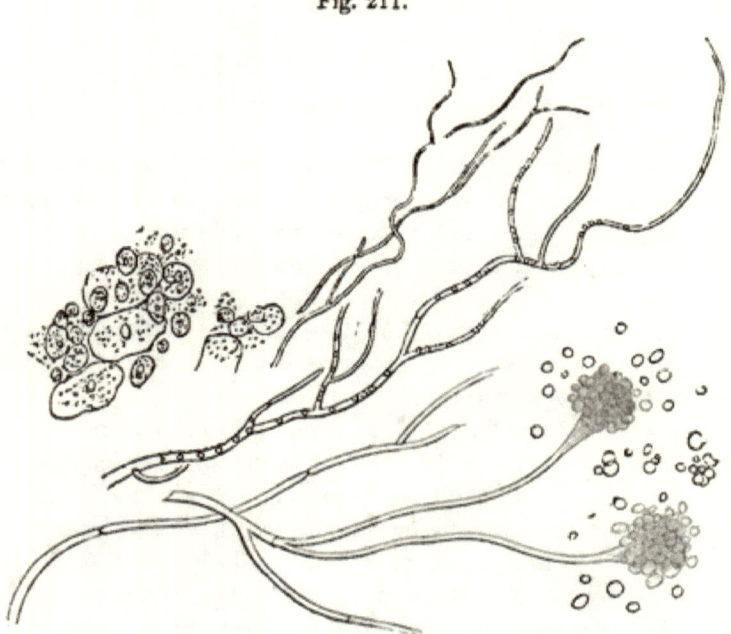

Oidium albicans from the mouth, with masses of epithelium.

Aspergillus.—A fungus of this species attacks the nail, penetrating its structure, and causing discoloration of the part, and leading to distortion.

Another small variety of *Aspergillus* has been found in the external auditory meatus; it consists of a long stem having a small cup, upon which numerous small spores are seated, as seen in the illustration.

FUNGI.

Fig. 212.

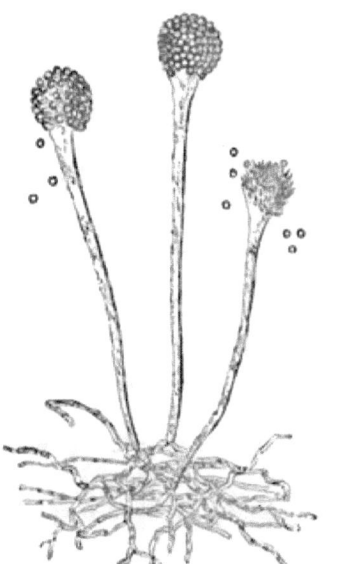

Aspergillus from the meatus auditorius.

Mycetoma Carteri, the foot-fungus of India. In this affection, the bones of the foot become perforated in every direction with canals, varying from the size of a pea to that of a nut, presenting the appearance indicated in the drawing, copied from the 'Intellectual Observer.'

Fig. 213.

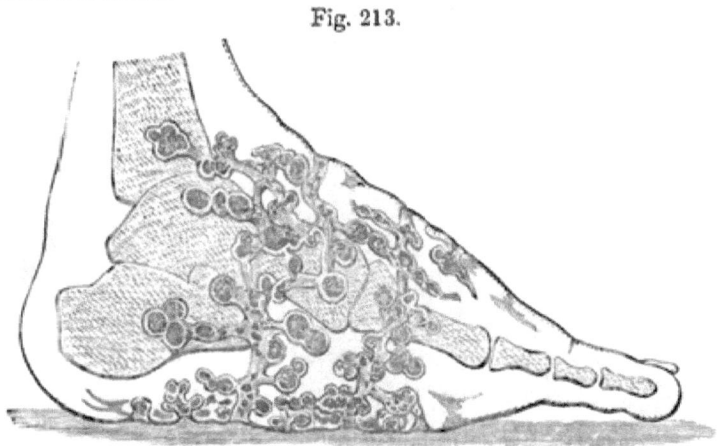

THE FOOT-FUNGUS OF INDIA.

The cavities are filled with a mass of fungus, red internally, but having an external coating of black. In its mature form the fungus consists of minute articulated threads, branched, and filled with grumous matter.

The spores contain an oil globule at either extremity, and germinate very rapidly.

Trichophyton tonsurans is formed of filaments arranged in rows. The spores originate within the filaments, and are seen as minute round bodies placed at short intervals.

The fungus penetrates the roots of the hair, spreading in an upward direction, causing small elevations upon the scalp. The hair ultimately breaks off at the epidermis.

The disease is termed *Herpes tonsurans*.

To obtain the fungus, it is necessary to pull out some of the diseased hairs, as it never occurs among the epidermoid scales of the scalp.

The annexed drawing represents a hair filled with the spores of *Trichophyton tonsurans*.

Fig. 214.

Hair, with the spores of Trichophyton tonsurans.

FUNGI.

Microsporon **andonini** surrounds the hair outside the follicle, instead of being developed at the root. Masses of it are sometimes heaped up around the openings of the follicles.

The fungus consists of numerous curved branches and filaments, to which small spores without any granules in their interior are found adhering.

The situation of the parasite suffices, however, to distinguish it from *Trichophyton tonsurans*.

Allopecia, or loss of hair, is the result of the growth of *Microsporon andonini*. The hairs become brittle and break off.

Microsporon mentagrophytes has larger filaments and branches, as well as larger spores, than the preceding.

The fungus attacks the hair of the beard, penetrating to the root. The hair becomes covered with yellowish and grey scales, its connection with the sheath is destroyed, it is rendered loose, and sometimes falls off spontaneously.

A drawing is given of this fungus below.

Fig. 215.

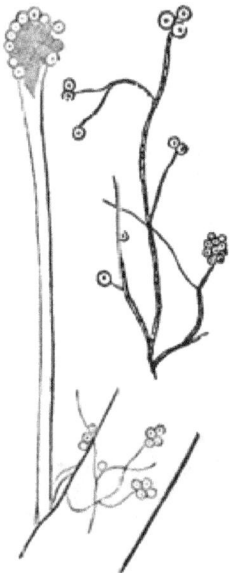

Microsporon mentagrophytes.

Microsporon furfur is associated with *Ptyriasis versicolor*, indicated by the formation of yellowish spots upon the skin of the chest and extremities. The spots increase from the size of a pea to the breadth of six or seven inches.

Under the microscope the fungus will be discovered in the upper horny layer of the epidermis of the affected parts. Acetic acid or caustic ammonia may be employed in the microscopic investigation.

The parasite is composed of long branched filamentous cells. The spores are often collected in groups, as shown in the illustration; they are highly refractive, and appear to be surrounded by two concentric lines.

Fig. 216.

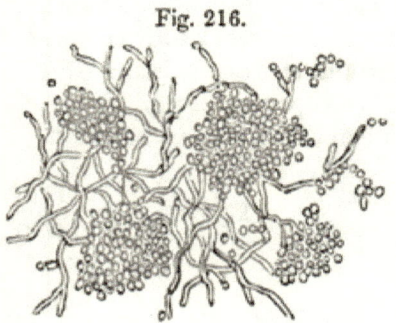

Microsporon furfur.

Achorion Schönleinii is found in favus, attaching itself to the bottom of the hair follicle, the spores being distributed over the root of the hair and upon the surface of the epidermis, round the favus crust. The bulb of the hair is often pointed, and frequently splits up into fibres.

Fig. 217.

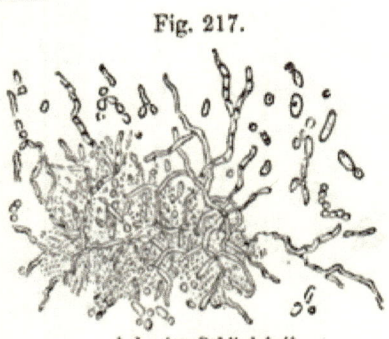

Achorion Schönleinii.

The mycelium of the fungus is formed of small tubes with partitions branched in all directions.

The spores are round or oval, and sometimes grouped together, as shown in the illustration. (Fig. 217.)

ALGÆ.

Cryptococcus cerevisiæ, Torula cerevisiæ, Yeast-plant. This plant is found in fermenting fluids, in yeast, and in saccharine urine. It occurs also in the mouth, œsophagus, and stomach.

It consists of numerous round or oval cells, containing in their interior minute granules resembling excessively minute oil globules.

The mode of propagation is by a species of budding; each cell gives forth one or two minute projections, which become perfect cells, and again develop other buds. In a few hours a row of five or six cells will be developed from a single one.

Fig. 218.

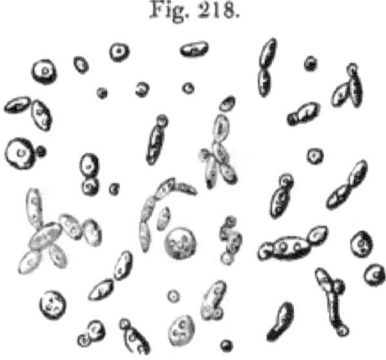

Torula cerevisiæ.

When found in any of the fluids, it indicates the existence of saccharine fermentation.

Sarcina ventriculi is composed of small square bodies, like wool-sacks, heaped together to form square masses, which consist of four, eight, sixteen, or sixty-four cubic cells (genidia).

Each cell is marked by cross lines or depressions, which

apparently divide it into four portions. Sometimes the cells are round, oval, or even triangular, but always with the corners rounded off.

The illustration represents several forms of the plant.

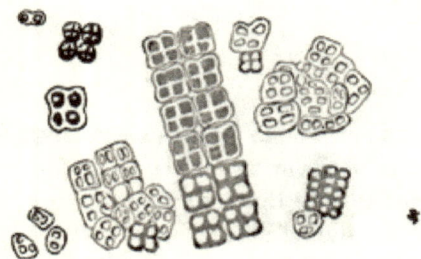

Fig. 219.

Various forms of Sarcina ventriculi.

Sarcina has been found in the human subject in the vomit, fæces, urine, pus of an abscess, and in the anterior chamber of the eye.

It may give rise to few or no symptoms; it may occur but once in the vomited matters, or, on the other hand, it may continue for weeks, and ultimately cause death from the constant vomiting it sometimes induces.

In order to discover it, the vomited matters should be allowed to rest for a time, and then both the serum on the surface and the sediment at the bottom of the liquid examined on a glass slide under a high power, as the little sac-like bodies are very minute, and may escape detection.

The fluid in which sarcina is found is usually fermenting, and sometimes it is of the consistence of pea-soup and has a brown or green appearance.

This species of alga is easily prepared by being put up in a drop of Canada balsam.

Leptothrix buccalis is found upon the tongue and in the masses of food which are sometimes allowed to collect between the teeth. The parasite is composed of fine filaments, sometimes attached to a stem. Under a very high power small granules may be seen in it.

From the healthy mouth the leptothrix may be obtained by

simply scraping the surface of the tongue, and examining the material collected in a drop of water under a high power. The algæ will be found either attached to masses of epithelium or lying free in the fluid.

Fig. 220.

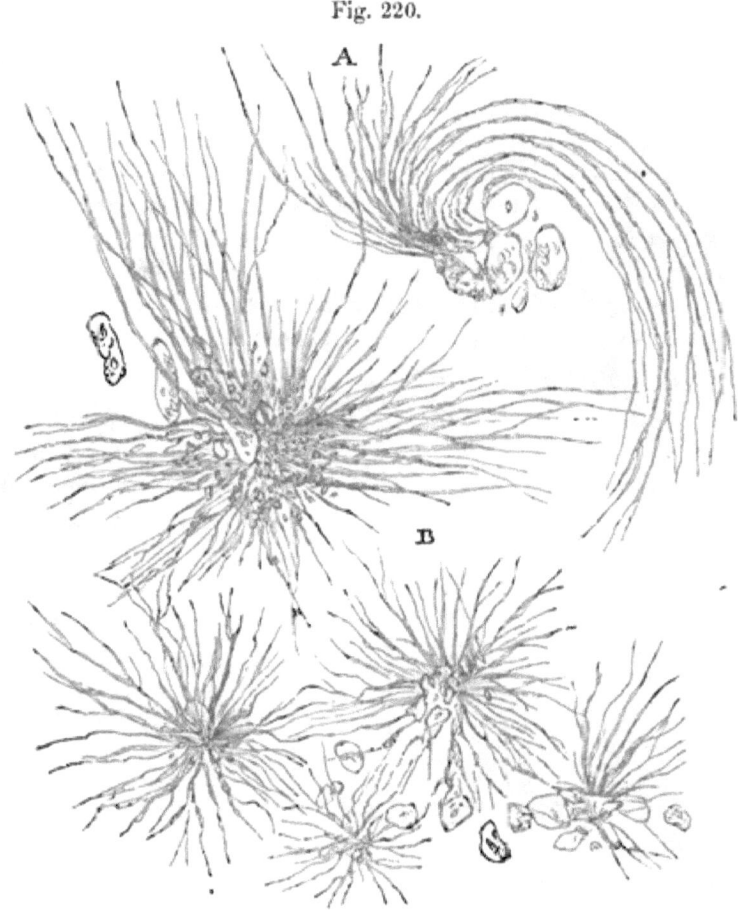

Algæ from the mouth and pharynx. A. Leptothrix buccalis from the tongue. B. Specimen in diphtheritic exudation.—Magnified 200 times.

The drawing shows the parasite from the healthy mouth, and also a specimen from the pharynx of a girl that died from diphtheria.

The mass of exudation which contained the specimens (B) was

of a greasy character, and the smaller tufts might have been mistaken for fatty acids. They were found in the tonsils, the pharynx, and also in the œsophagus, and were not fairly brought into view until alcohol was added.

From the healthy tongue the *Leptothrix buccalis* may be obtained in abundance; in the morning particularly a portion of the yellowish furry material from the centre of the organ will usually afford good specimens. A high power must be used in the examination, as the filaments are exceedingly delicate.

In the course of his inquiries the microscopist will meet with numerous forms of fungi and algæ differing from those which have deen described. Indeed it is yet uncertain to what extent new varieties of parasitic plants may arise in consequence of an altered condition of the secretions; but it does not appear in the present state of our knowledge that these new forms possess any special pathological significance.

INDEX.

A CHORION Schönleinii, 258
Acari, 223—225
Adenoid tumours, 212
Adipose tissue, 54
Adventitious products, 195
   osseous tumours, 196
      exostosis, 196
   osteophytes, 198
   osteoma, 199
   osteosarcoma, 199
   enchondroma, 199
      section of, 200
   fibrous tumours, 200
      recurrent fibroid, 201
   indurated chancre, 202
   fatty tumours (lipomata), 202
      section of 'reticulated fatty tumour' on a child's foot, 204
   myeloid growths, 205
   hæmatoma, 205, 206
Air-bubbles, 20
Algæ, 259—261
Allopecia, causes of, 257
Anchylostomum duodenale, 247
Angeiectoma, 205
Aorta of the horse, 139
Apparatus required for the preparation of the objects for examination, 11
   bull's-eye condenser, 12
   side reflector, 12
   lieberkühn, 12
   dark wells or stops, 12
   achromatic condenser, 12
   camera lucida, 13
      1. Wollaston's prism, 13
      2. Sœmmering's speculum, 13
      3. Neutral glass reflector, 13
   micrometer, 13
   parabolic reflector, 14
   polariscope, 14
   erector, 15
   microscope lamp, 15

Apparatus—*continued.*
   Valentin's knife, 15
   compressorium, 15
   animalcule cage, 15
   zoophyte trough, 16
   frog plate, 16
   stage forceps, 16
Areolar tissue, 51
Arteries, preparations of the, for the microscope, 138
   aorta of the horse, 139
      various coats of which it is composed, 140
Ascarides, 239
Aspergillus, 234

BACILLAR layer, or Membrana Jacobi, 159
Bacteria and blood, 184
Basement membrane, 49
Biliary calculi, 185
Blood, microscopic examination of the, 184
   bacteria and vibriones, 184
Blood-cells, 39
   specimens for examination, 39
   characters of a single red disc, 41
   sizes of blood discs in various animals, 41
   number of red corpuscles in a given space, 41
   white corpuscles, 42
   blood discs of the fowl, 43
      in fishes and in reptiles, 44
   changes in the blood-cells by reagents, 44
   observation of the blood in circulation, 46
Blood-crystals, 47, 48
   hæmatin crystals, found in normal blood, 47
   hæmatoidin crystals, 47
   hæmin crystals, 48

Bone, structure of, 62
  microscopic examination of, 62
  development of bone, 67
    ossification of cartilage, 67
      examination of ossifying cartilage, 68
Brain, the, 150

CALCULI. *See* Concretions.
Cancer, improper use of the term, 217
  See Encephaloid growths
  secondary cancer, 220
Canker of the foot of a horse, 207, 208
Capillaries, structure of, 140, 141
  capillaries of the human brain, 141
Cartilage, or gristle, structure of, 56
  permanent and temporary, 56
  cellular cartilage, 59
  yellow fibro-cartilage, 60
  white fibro-cartilage, 61
Cartilage-cells, human, 28
  of a tadpole, 30
Cartilago nictitans, 151
Cataract, black, 193
Cells, their nature and functions, 27
  simple cells, 27
    primordial utricle, 28
    preparation, 28
  origin of cells, 29
  on the function of cells, 31
    simple cells forming a covering, 31
    cells performing the office of a storehouse, 31
      preparation, 31
  complex cells, 33
  cells with moving contents, 34
    preparation of Vallisneria and Vitella, 34
  epithelial cells, 35
    tesselated epithelium, 35
    columnar epithelium, 36
      preparation, 36
    spheroidal epithelium, 37
      preparation, 37
    ciliated epithelium, 37
      preparation, 38
  lymph, 38
  chyle, 39
  blood, 39
Cestoda, 229—237
Chancre, indurated, 202
Choroid coat, 153
  vessels of the, 156
Chyle, 39
Colloid tumours, 215
  of the ovary, 215
Concretions, microscopic examination of, 185
  biliary calculi, 185

Concretions—*continued.*
  intestinal concretions, 186
  animal concretions, 187
  hair concretions, 187
  oat-hair calculi, 188
  starch, 188
  urinary, 188
Conjunctiva, the, 152
Cornea, section of the, 152
Corpora lutea, 173, 174
Corpora nigra, 157
Crab-louse, 227
Cryptococcus cerevisiæ, 239
Crystalline lens, 160, 161
Cysticercus cellulosæ, 231
Cystoid tumours, 213
  sebaceous cysts, 213
  glandular cysts, 213
  synovial cysts, 213
  compound cysts, 214

DENTAL pulp, the, 107
  Dentine, 104
Digestive canal, 108
  racemose glands from the mouth, 108
  follicular gland from the root of the tongue, 108
  upper surface of the human tongue, 109
  papillæ of the tongue, 110
  salivary glands, 111
    parotid, 111
  mucous membrane of the œsophagus, 112
  mucous membrane of the stomach, 112
  mucous membrane of the small intestines, 114
Diphtheria, leptothrix buccalis in a case of, 261
Distoma, 248—252

EAR, the internal, 164
  transverse section of the lamina spiralis, 164
  methods of preparation, 165
Elementary tissues, 27
Elephantiasis, longitudinal and transverse sections, 210
Enamel of the teeth, 106
Encephaloid growths, 218
  elements of an encephaloid tumour, 219
    characteristics of malignant disease, 219
Enchondroma, 199
Epidermoid growths, 206
  canker, foot-rot, and thrush, 207
  disease of the nails, 208, 209
  elephantiasis, 210

Epithelial cells. *See* Cells
Epithelial growths, 211
   epithelial cancer, 211
Epizoa, 222—225
Eye, the, 151
   the cartilago nictitans, 151
   preparations for examination, 152, 153
   section of the cornea and sclerotic, 152
   the conjunctiva and membrana humoris aquei, 153
   nerves of the cornea of the rabbit in their coarser ramifications, 154
   the choroid coat, 154
   the tapetum lucidum, 155
   vessels of the choroid and iris, 156
   nerves of the iris, 157
   corpora nigra, 157
   retina, 157
      ora serrata, 157
      pars ciliaris retinæ, 157
      vertical section of the human retina, 158
   the bacillar layer, or membrana jacobi, 159
   nerve-cells from the human retina, 160
   the crystalline lens, 160
      lenticular tubes or fibres, 161
   the vitreous body, 163
Exostosis, 196

FAT. See Tissue, adipose
   fatty tumours, 202
Fibrous tissue, 49
Filariæ, 241—243
Fluke-shaped worms, 248
Foot-rot of sheep, 207
Foot-fungus of India, 255
Fungi, 253—258

GENERATION, organs of, 168
   seminal tubule of man, 168
      development of the spermatic filaments of a bull, 169
      human spermatozoa, 170
      spermatozoa from the insect tribe, 170
         from some fishes, 171
         from the cock, 171
         from sparrows and finches, 171
         from rats and mice, 171, 172
   ovaries of the female, 172
      section of an ovary, 172
      human ovulum, 173
      corpora lutea, 173, 174
      mammary glands, 174
      microscopic examination of milk, 175

HÆMATIN crystals in normal blood, 47
   Hæmatoidin crystals, 47
Hæmatoma, 205, 206
Hæmin crystals, 48
Hair, structure of, 90
   microscopic examination of, 91
Hair concretions, 187
Heart, fatty degeneration of the, 190
Herpes tonsurans, 256
Hoof, structure of, 100
   transverse section of horse's, 101
Horn, structure of, 100

INTEGUMENT, structure of the, 78
   pigment-cells in the skin of the negro, 79
   dermis, cutis vera, or true skin, 80
   papillæ, 81
   nerves of the skin, 83
   subcutaneous cellular tissue, 83
   cutaneous glands, 84
      sudoriferous, 84
      sebaceous, 86
      ceruminous, 88
   bronzed skin, 193—195
   sections of integument in elephantiasis, 210
Intestines, mucous membrane of the small, 114
Intestines, concretions in the, 186
Iris Germanica, section of root of, 32
   cuticle of, showing stomata, 34
Iris, vessels of the, 153
   nerves of the, 157

KIDNEYS, characteristics of each of the, 129
   preparations for the microscope, 130
   relations of the various parts, 132
   transverse sections, 133

LAMINA spiralis, transverse section of the, 164
Leptothrix buccalis, 260, 261
Lice, from head, body, and pubis of human subject, 227
Lime, oxalate of, in urine, 183
Lipomata, 202
Lipomata arborescentia, 203
Liver, structure of the, 121
   hepatic cells, 121, 122
   fatty degeneration of the, 190, 191.
Livers of Strasburg geese, 189
Lungs, structure of the, 135
   preparations for the microscope, 135
   pulmonary vesicle with the parts adjoining, 136

Lungs—*continued.*
  pulmonary lobules, 136, 137
  distribution of the bloodvessels in relation to the pulmonary vesicles, 137
  tuberculous deposit in the, 192
Lymph, 39
Lymphatics, structure of the, 142

MAMMARY glands, 174
  schirrous tumour of the, 216
Melanine, 196
Melanotic infiltration, 194—196
Membrana humoris aquei, 153
Membrana jacobi, or bacillar layer, 159
Microscope, description of the, 1
  simple microscope, 1
  compound microscope, 2, 7
  refraction, 2
  reflection, 2
  various kinds of lenses, 3
  spherical and chromatic aberration,
  binocular microscope, 9
  additional apparatus, 12
  mode of using the microscope, 16
  focussing, 21
  preserving and mounting objects for the microscope, 22
    preservative fluids, 22
    colouring fluids, 22
    injected preparations, 25
Microsporon andouini, 257
  mentagrophytes, 257
  furfur, 258
Milk, microscopic examination of, 175
  in a morbid state, 182
Morbid histology, 177 *et seq.*
  microscopic examination of morbid fluids, 179
Mucous secretion, 167
Muscle, structure of, 70
  examination of striated or voluntary muscle, 70
    examination of the fibres, 71
    examination of fibrillæ, 74
  involuntary muscle, 76
    striated and non-striated varieties, 76
  diseased muscular tissue, 189
Mycetoma Carteri, 255
Myeloid growths, 205

NAILS, structure of the, 97
  microscopic examination of, 97
  time of growth of, 99
  peculiar forms of disease affecting the nails of the human subject, 208, 209
Nematoda, 239

Nervous system, the, 142
  preparations for the microscope, 142
  examination of nervous fibrillæ, 143
  gelatinous or grey fibres, 145
  nerve-cells or vesicles, 147
    granular corpuscles surrounding ganglia cells, 147
    preparation of caudate cells, 148
  Pacinian bodies, 148
  the brain and spinal cord, 150
  Mr. Lockhart Clarke's method of making preparations of nerve-structure, 150
Nose, structure of the, 165
  the epithelial cells, 165
  sections of the mucous membrane, 165, 166
  olfactory nerves, 166, 167
  mucous secretion, 167

OAT-HAIR calculi, 188
  Œsophagus, mucous membrane of the, 112
Oidium albicans, 253
Oil-globules, 20
Olfactory nerves, 166, 167
Onion, membrane from inner layers of an, 31
Osseous tumours, 196
Osteo-dentine, 106
Osteoma, 197
Osteophytes, 198
Osteosarcoma, 199
Ova of Ascaris nigrovenosa, 20
Ovaries, 172
  section of an ovary, 172
  human ovulum, 173
  colloid tumours of the, 215
Oxyuris vermicularis, 241

PACINIAN bodies, 148
  Pancreas, structure of the, 120
    microscopic examination of the, 120
Parasites, animal, 221
  preparations of, 221, 222
  epizoa, 222
    Acarus folliculorum, 223
    Acarus scabiei, 224, 225
    Acarus ovis, 224, 225
    ticks, 225
    Pediculi, 225
  entozoa, 228
    Cestoda—tape-worms, 229
      Tænia solium, 230
        Cysticercus cellulosæ, 231
      Tænia mediocanellata, 232
      Tænia marginata, 233
        Cysticercus tenuicollis, 233

INDEX.

Parasites, animal—*continued*.
    Tænia echinococcus, 234
    Tænia serrata, 235
        Cysticercus pisiformis, 235
    Tænia crassicollis, 235
        Cysticercus fasciolaris, 236
    Tænia cænurus, 236
    Tænia elliptica, 237
    Tænia cucumerina, 237
    Bothriocephalus latus, 237
    Nematoda—round worms, 239
        Ascaris lumbricoides, 239
        Ascaris mystax, 239
        Ascaris vermicularis, 239
        Ascaris megalocephala, 239, 240
        Oxyuris vermicularis, 241
        Filaria bronchialis, or Strongylus bronchialis, 241
        Filaria oculi, 243
        Filaria Medinensis, or Guinea worm, 243
        Strongylus gigas, 244
        Strongylus paradoxus, 244
        Trichocephalus dispar, 244
        Trichina spiralis, 245
            cysts of Trichinæ in situ, 246
            mature Trichina found in the intestines, 247
        Sclerostoma syngamus, 247
        Sclerostoma duodenale, or Anchylostomum duodenale, 247
        Pentastoma, 248
            tænoides, 248
            denticulatum, 248
    Trematoda—fluke-shaped worms, 248
        Distoma hepaticum, or Fasciola hepaticum, 248, 249
            the several stages of development, 250
        Distoma lanceolatum, 251
        Distoma opthalmobium, 251
        Distoma hæmatobium, 251
        Distoma heterophyes, 252
Parasites, vegetable, 252
    method of examination, 253
    fungi, 253
        Oidium albicans, 253
        Aspergillus, 254
        Mycetoma Carteri, 255
        Trichophyton tonsurans, 256
        Microsporon audonini, 257
        Microsporon mentagrophytes, 257
        Microsporon furfur, 258
            Achorion Schönleinii, 258
    algæ, 259
        Cryptococcus cerevisiæ, Torula cerevisiæ, yeast-plant, 259
        Sarcina ventriculi, 259
        Leptothrix buccalis, 260, 261

Parotid glands, 111
Pentastoma, 248
Phosphate, crystals of triple, in urine, 183
Pigment, black, in man, cells of, 33
Pigmentary degeneration, 193
Pus, microscopic examination of, 179

RETINA, 157
    ora serrata, 157
    pars ciliaris retinæ, 157
    nerve-cells, 160
Round worms, 239

SALIVARY glands, 111
    Sarcina ventriculi, 259
Schirrous growths, 216
    schirrous tumour of the mammary gland, 216
Sclerostoma syngamus, 247
    duodenale, 247
Sclerotic, section of the, 152
Seedy toe of the horse, 209, 210
Serous fluid, microscopic examination of, 181
Skin, the. *See* Integument.
Spermatozoa, 169
Spermatozoa, human, 170
    from insects and from various animals, 170—172
Spinal cord, the, 150
Spleen, structure of the, 127
    cells, blood-discs, &c., 128
Starch-granules from a potato, 32
Starch calculi, 188
Stomach, mucous membrane of the, 112
Strongylus gigas, 244
    paradoxus, 244
Supra-renal capsules, 133
    transverse section, 133

TADPOLE, cartilage-cells of a, 30
    Tæniæ, 230—237
Tapetum lucidum, 153
Tape-worms, 229—237
Teeth, structure of, 102
    sections of a human molar tooth, 102
    section of incisor of a horse, 103
    microscopic examination of tooth, 103
        dentine, 104
        osteo-dentine, 106
        enamel, 106
        dental pulp, 107
Textures, degeneration of, 189
    diseased livers of Strasburg geese, 189
    diseased muscular tissue, 189

Textures—*continued.*
    fatty degeneration of the heart, 190
    fatty degeneration of the liver, 190, 191
    calcareous degeneration, 191
    tuberculous deposit, 191
        in the lungs, 192
    typhous matter, 193
    pigmentary degeneration, 193
        black cataract, 193
    bronzed skin, 194, 195
Thrush of the horse's foot, 207
Thrush-fungus, 253
Thymus gland, structure of the, 124
    preparation of specimens, 125, 126
Thyroid gland, structure of the, 123
    preparation of specimens, 123
Ticks, 223—225
Tissue, adipose, 54
    areolar, 51
    fibrous, 49
    white, 49
    yellow or elastic, 49
Tongue, upper surface of the human, 109
    papillæ of the, 110

Torula cerevisiæ, 259
Trematoda, 248
Trichina spiralis, 245—247
Trichocephalus dispar, 244
Trichophyton tonsurans, 256
Tumours, osseous, 196
    fibrous, 200
    fatty, 202
Typhous matter, 193

URIC acid, crystals of, 182
    Urine, microscopic examination of, 182
    urinary deposits, 182
        crystals of uric acid, 182
        oxalate of lime, 183
        crystals of triple phosphate, 183
        organic products, 184
    urinary calculi, 188

VEINS, structure of, 140
    Vibriones in blood, 184
Vitreous body of the eye, 163

[MARCH 1866.]

GENERAL LIST OF WORKS

PUBLISHED BY

Messrs. LONGMANS, GREEN, AND CO.

PATERNOSTER ROW, LONDON.

Historical Works.

LORD MACAULAY'S WORKS. Complete and Uniform Library Edition. Edited by his Sister, Lady TREVELYAN. 8 vols. 8vo. with Portrait, price £5 5s. cloth, or £8 8s. bound in tree-calf by Rivière.

The HISTORY of ENGLAND from the Fall of Wolsey to the Death of Elizabeth. By JAMES ANTHONY FROUDE, M.A. late Fellow of Exeter College, Oxford.

    VOLS. I. to IV. the Reign of Henry VIII. Third Edition, 54s.
    VOLS. V. and VI. the Reigns of Edward VI. and Mary. Second Edition, 28s.
    VOLS. VII. and VIII. the Reign of Elizabeth, VOLS. I. and II. Third Edition, 28s.

The HISTORY of ENGLAND from the Accession of James II. By Lord MACAULAY.

    LIBRARY EDITION, 5 vols. 8vo. £4.
    CABINET EDITION, 8 vols. post 8vo. 48s.
    PEOPLE'S EDITION, 4 vols. crown 8vo. 16s.

REVOLUTIONS in ENGLISH HISTORY. By ROBERT VAUGHAN, D.D. 3 vols. 8vo. 45s.

    VOL. I. Revolutions of Race, Second Edition, revised, 15s.
    VOL. II. Revolutions in Religion, 15s.
    VOL. III. Revolutions in Government, 15s.

An ESSAY on the HISTORY of the ENGLISH GOVERNMENT and Constitution, from the Reign of Henry VII. to the Present Time. By JOHN EARL RUSSELL. Third Edition, revised. Crown 8vo. 6s.

The HISTORY of ENGLAND during the Reign of George the Third. By the Right Hon. W. N. MASSEY. Cabinet Edition. 4 vols. post 8vo. 24s.

The CONSTITUTIONAL HISTORY of ENGLAND, since the Accession of George III. 1760—1860. By THOMAS ERSKINE MAY, C.B. Second Edition. 2 vols. 8vo. 33s.

A

CONSTITUTIONAL HISTORY of the BRITISH EMPIRE from the Accession of Charles I. to the Restoration. By G. BRODIE, Esq. Historiographer-Royal of Scotland. Second Edition. 3 vols. 8vo. 36s.

HISTORICAL STUDIES. I. On Some of the Precursors of the French Revolution; II. Studies from the History of the Seventeenth Century; III. Leisure Hours of a Tourist. By HERMAN MERIVALE, M.A. 8vo. price 12s. 6d.

LECTURES on the HISTORY of ENGLAND. By WILLIAM LONGMAN. VOL. I. from the earliest times to the Death of King Edward II. with 6 Maps, a coloured Plate, and 53 Woodcuts. 8vo. 15s.

HISTORY of CIVILISATION. By HENRY THOMAS BUCKLE. 2 vols. 8vo. £1 17s.
   VOL. I. *England and France*, Fourth Edition, 21s.
   VOL. II. *Spain and Scotland*, Second Edition, 16s.

DEMOCRACY in AMERICA. By ALEXIS DE TOCQUEVILLE. Translated by HENRY REEVE, with an Introductory Notice by the Translator. 2 vols. 8vo. 21s.

The SPANISH CONQUEST in AMERICA, and its Relation to the History of Slavery and to the Government of Colonies. By ARTHUR HELPS. 4 vols. 8vo. £3. VOLS. I. and II. 28s. VOLS. III. and IV. 16s. each.

HISTORY of the REFORMATION in EUROPE in the Time of Calvin. By J. H. MERLE D'AUBIGNÉ, D.D. VOLS. I. and II. 8vo. 28s. and VOL. III. 12s. VOL. IV. nearly ready.

LIBRARY HISTORY of FRANCE, in 5 vols. 8vo. By EYRE EVANS CROWE. VOL. I. 14s. VOL. II. 15s. VOL. III. 18s. VOL. IV. nearly ready.

LECTURES on the HISTORY of FRANCE. By the late Sir JAMES STEPHEN, LL.D. 2 vols. 8vo. 24s.

The HISTORY of GREECE. By C. THIRLWALL, D.D. Lord Bishop of St. David's. 8 vols. 8vo. £3; or in 8 vols. fcp. 28s.

The TALE of the GREAT PERSIAN WAR, from the Histories of Herodotus. By GEORGE W. COX, M.A. late Scholar of Trin. Coll. Oxon. Fcp. 7s. 6d.

GREEK HISTORY from Themistocles to Alexander, in a Series of Lives from Plutarch. Revised and arranged by A. H. CLOUGH. Fcp. with 44 Woodcuts. 6s.

CRITICAL HISTORY of the LANGUAGE and LITERATURE of Ancient Greece. By WILLIAM MURE, of Caldwell. 5 vols. 8vo. £3 9s.

HISTORY of the LITERATURE of ANCIENT GREECE. By Professor K. O. MÜLLER. Translated by the Right Hon. Sir GEORGE CORNEWALL LEWIS, Bart. and by J. W. DONALDSON, D.D. 3 vols. 8vo. 36s.

HISTORY of the CITY of ROME from its Foundation to the Sixteenth Century of the Christian Era. By THOMAS H. DYER, LL.D. 8vo. with 2 Maps, 15s.

HISTORY of the ROMANS under the EMPIRE. By CHARLES MERIVALE, B.D. Chaplain to the Speaker. Cabinet Edition, with Maps, complete in 8 vols. post 8vo. 48s.

The FALL of the ROMAN REPUBLIC: a Short History of the Last Century of the Commonwealth. By CHARLES MERIVALE, B.D. Chaplain to the Speaker. Fourth Edition. 12mo. 7s. 6d.

The CONVERSION of the ROMAN EMPIRE: the Boyle Lectures for the year 1864, delivered at the Chapel Royal, Whitehall. By CHARLES MERIVALE, B.D. Chaplain to the Speaker. Second Edition, 8vo. 8s. 6d.

The CONVERSION of the NORTHERN NATIONS; the Boyle Lectures for 1865. By the same Author. 8vo. 8s. 6d.

CRITICAL and HISTORICAL ESSAYS contributed to the *Edinburgh Review*. By the Right Hon. LORD MACAULAY.

> LIBRARY EDITION, 3 vols. 8vo. 36s.
> TRAVELLER'S EDITION, in 1 vol. 21s.
> CABINET EDITION, 3 vols. fcp. 21s.
> PEOPLE'S EDITION, 2 vols. crown 8vo. 8s.

HISTORICAL and PHILOSOPHICAL ESSAYS. By NASSAU W. SENIOR. 2 vols. post 8vo. 16s.

HISTORY of the RISE and INFLUENCE of the SPIRIT of RATIONALISM in EUROPE. By W. E. H. LECKY, M.A. Second Edition, revised. 2 vols. 8vo. 25s.

The HISTORY of PHILOSOPHY, from Thales to the Present Day. By GEORGE HENRY LEWES. Third Edition, partly rewritten and greatly enlarged. In 2 vols. VOL. I. *Ancient Philosophy*; VOL. II. *Modern Philosophy*. [*Nearly ready.*

HISTORY of the INDUCTIVE SCIENCES. By WILLIAM WHEWELL, D.D. F.R.S. late Master of Trin. Coll. Cantab. Third Edition. 3 vols. crown 8vo. 24s.

HISTORY of SCIENTIFIC IDEAS; being the First Part of the Philosophy of the Inductive Sciences. By the same Author. 2 vols. cr. 8vo. 14s.

EGYPT'S PLACE in UNIVERSAL HISTORY; an Historical Investigation. By C. C. J. BUNSEN, D.D. Translated by C. H. COTTRELL, M.A. With many Illustrations. 4 vols. 8vo. £5 8s. VOL. V. is nearly ready.

MAUNDER'S HISTORICAL TREASURY; comprising a General Introductory Outline of Universal History, and a series of Separate Histories. Fcp. 10s.

HISTORICAL and CHRONOLOGICAL ENCYCLOPÆDIA, presenting in a brief and convenient form Chronological Notices of all the Great Events of Universal History. By B. B. WOODWARD, F.S.A. Librarian to the Queen. [*In the press.*

HISTORY of the CHRISTIAN CHURCH, from the Ascension of Christ to the Conversion of Constantine. By E. BURTON, D.D. late Prof. of Divinity in the Univ. of Oxford. Eighth Edition. Fcp. 3s. 6d.

SKETCH of the HISTORY of the CHURCH of ENGLAND to the Revolution of 1688. By the Right Rev. T. V. SHORT, D.D. Lord Bishop of St. Asaph. Seventh Edition. Crown 8vo. 10s. 6d.

HISTORY of the EARLY CHURCH, from the First Preaching of the Gospel to the Council of Nicæa, A.D. 325. By the Author of 'Amy Herbert.' Fcp. 4s. 6d.

The **ENGLISH REFORMATION.** By F. C. MASSINGBERD, M.A. Chancellor of Lincoln and Rector of South Ormsby. Fourth Edition, revised. Fcp. 8vo. [*Nearly ready.*

HISTORY of WESLEYAN METHODISM. By GEORGE SMITH, F.A.S. Fourth Edition, with numerous Portraits. 3 vols. cr. 8vo. 7s. each.

LECTURES on the HISTORY of MODERN MUSIC, delivered at the Royal Institution. By JOHN HULLAH. FIRST COURSE, with Chronological Tables, post 8vo. 6s. 6d. SECOND COURSE, on the Transition Period, with 40 Specimens, 8vo. 16s.

Biography and *Memoirs.*

EXTRACTS of the JOURNALS and CORRESPONDENCE of MISS BERRY, from the Year 1783 to 1852. Edited by Lady THERESA LEWIS. Second Edition, with 3 Portraits. 3 vols. 8vo. 42s.

The DIARY of the Right Hon. WILLIAM WINDHAM, M.P. From 1783 to 1809. Edited by Mrs. HENRY BARING. 8vo. 18s.

LIFE of the DUKE of WELLINGTON. By the Rev. G. R. GLEIG, M.A. Popular Edition, carefully revised; with copious Additions. Crown 8vo. with Portrait, 5s.

Brialmont and Gleig's **Life of the Duke of Wellington.** (The Parent Work.) 4 vols. 8vo. with Illustrations, £2 14s.

Life of the Duke of Wellington, Intermediate Edition, partly from the French of M. BRIALMONT, partly from Original Documents. By the Rev. G. R. GLEIG, M.A. 8vo. with Portrait, 15s.

HISTORY of MY RELIGIOUS OPINIONS. By J. H. NEWMAN, D.D. Being the Substance of Apologia pro Vitâ Suâ. Post 8vo. 6s.

FATHER MATHEW: a Biography. By JOHN FRANCIS MAGUIRE, M.P. Popular Edition, with Portrait. Crown 8vo. 3s. 6d.

Rome; its Rulers and its Institutions. By the same Author. New Edition in preparation.

LIFE of AMELIA WILHELMINA SIEVEKING, from the German. Edited, with the Author's sanction, by CATHERINE WINKWORTH. Post 8vo. with Portrait, 12s.

MOZART'S LETTERS (1769-1791), translated from the Collection of Dr. LUDWIG NOHL by Lady WALLACE. 2 vols. post 8vo. with Portrait and Facsimile, 18s.

BEETHOVEN'S LETTERS (1790-1826), from the Two Collections of Drs. NOHL and discovered Letters to the Archduke Rudolph. Cardinal-Archbishop of Olmütz, VON KÖCHEL. Translated by Lady WALLACE. 2 vols. post 8vo. with Portrait.

FELIX MENDELSSOHN'S LETTERS from *Italy and Switzerland,* and *Letters from* 1833 *to* 1847, translated by Lady WALLACE. New Edition, with Portrait. 2 vols. crown 8vo. 5s. each.

RECOLLECTIONS of the late **WILLIAM WILBERFORCE, M.P.** for the County of York during nearly 30 Years. By J. S. HARFORD, F.R.S. Second Edition. Post 8vo. 7s.

MEMOIRS of SIR HENRY HAVELOCK, K.C.B. By JOHN CLARK MARSHMAN. Second Edition. 8vo. with Portrait, 12s. 6d.

THOMAS MOORE'S MEMOIRS, JOURNAL, and CORRESPONDENCE. Edited and abridged from the First Edition by Earl RUSSELL. Square crown 8vo. with 8 Portraits, 12s. 6d.

MEMOIR of the Rev. SYDNEY SMITH. By his Daughter, Lady HOLLAND. With a Selection from his Letters, edited by Mrs. AUSTIN. 2 vols. 8vo. 28s.

VICISSITUDES of FAMILIES. By Sir BERNARD BURKE, Ulster King of Arms. FIRST, SECOND, and THIRD SERIES. 3 vols. crown 8vo. 12s. 6d. each.

ESSAYS in ECCLESIASTICAL BIOGRAPHY. By the Right Hon. Sir J. STEPHEN, LL.D. Fourth Edition. 8vo. 14s.

BIOGRAPHIES of DISTINGUISHED SCIENTIFIC MEN. By FRANÇOIS ARAGO. Translated by Admiral W. H. SMYTH, F.R.S. the Rev. B. POWELL, M.A. and R. GRANT, M.A. 8vo. 18s.

MAUNDER'S BIOGRAPHICAL TREASURY: Memoirs, Sketches, and Brief Notices of above 12,000 Eminent Persons of All Ages and Nations. Fcp. 10s.

LETTERS and LIFE of FRANCIS BACON, including all his Occasional Works. Collected and edited, with a Commentary, by J. SPEDDING, Trin. Coll. Cantab. VOLS. I. and II. 8vo. 24s.

Criticism, Philosophy, Polity, &c.

The INSTITUTES of JUSTINIAN; with English Introduction, Translation, and Notes. By T. C. SANDARS, M.A. Barrister, late Fellow of Oriel Coll. Oxon. Third Edition. 8vo. 15s.

The ETHICS of ARISTOTLE. Illustrated with Essays and Notes. By Sir A. GRANT, Bart. M.A. LL.D. Director of Public Instruction in the Bombay Presidency. Second Edition, revised and completed. 2 vols. 8vo.

ELEMENTS of LOGIC. By R. WHATELY, D.D. late Archbishop of Dublin. Ninth Edition. 8vo. 10s. 6d. crown 8vo. 4s. 6d.

Elements of Rhetoric. By the same Author. Seventh Edition. 8vo. 10s. 6d. crown 8vo. 4s. 6d.

English Synonymes. Edited by Archbishop WHATELY. 5th Edition. Fcp. 3s.

BACON'S ESSAYS with ANNOTATIONS. By R. WHATELY, D.D. late Archbishop of Dublin. Sixth Edition. 8vo. 10s. 6d.

LORD BACON'S WORKS, collected and edited by R. L. ELLIS, M.A. J. SPEDDING, M.A. and D. D. HEATH. Vols. I. to V. *Philosophical Works,* 5 vols. 8vo. £4 6s. VOLS. VI. and VII. *Literary and Professional Works,* 2 vols. £1 16s.

On **REPRESENTATIVE GOVERNMENT.** By JOHN STUART MILL, M.P. for Westminster. Third Edition, 8vo. 9s. crown 8vo. 2s.

On **Liberty.** By the same Author. Third Edition. Post 8vo. 7s. 6d. crown 8vo. 1s. 4d.

Principles of Political Economy. By the same. Sixth Edition. 2 vols. 8vo. 30s. or in 1 vol. crown 8vo. 5s.

A System of Logic, Ratiocinative and Inductive. By the same. Sixth Edition. Two vols. 8vo. 25s.

Utilitarianism. By the same. Second Edition. 8vo. 5s.

Dissertations and Discussions. By the same Author. 2 vols. 8vo. price 24s.

Examination of Sir W. Hamilton's Philosophy, and of the Principal Philosophical Question discussed in his Writings. By the same Author. Second Edition. 8vo. 14s.

MISCELLANEOUS REMAINS from the Common-place Book of RICHARD WHATELY, D.D. late Archbishop of Dublin. Edited by Miss E. J. WHATELY. Crown 8vo. 7s. 6d.

ESSAYS on the ADMINISTRATIONS of GREAT BRITAIN from 1783 to 1830. By the Right Hon. Sir G. C. LEWIS, Bart. Edited by the Right Hon. Sir E. HEAD, Bart. 8vo. with Portrait, 15s.

By the same Author.

Inquiry into the Credibility of the Early Roman History, 2 vols. price 30s.

On the Methods of Observation and Reasoning in Politics, 2 vols. price 28s.

Irish Disturbances and Irish Church Question, 12s.

Remarks on the Use and Abuse of some Political Terms, 9s.

The Fables of Babrius, Greek Text with Latin Notes, PART I. 5s. 6d. PART II. 3s. 6d.

An OUTLINE of the NECESSARY LAWS of THOUGHT: a Treatise on Pure and Applied Logic. By the Most Rev. W. THOMSON, D.D. Archbishop of York. Crown 8vo. 5s. 6d.

The ELEMENTS of LOGIC. By THOMAS SHEDDEN, M.A. of St. Peter's Coll. Cantab. 12mo. 4s. 6d.

ANALYSIS of Mr. MILL'S SYSTEM of LOGIC. By W. STEBBING, M.A. Fellow of Worcester College, Oxford. Second Edition. 12mo. 3s. 6d.

The ELECTION of REPRESENTATIVES, Parliamentary and Municipal; a Treatise. By THOMAS HARE, Barrister-at-Law. Third Edition, with Additions. Crown 8vo. 6s.

SPEECHES of the RIGHT HON. LORD MACAULAY, corrected by Himself. Library Edition, 8vo. 12s. People's Edition, crown 8vo. 3s. 6d.

LORD MACAULAY'S SPEECHES on PARLIAMENTARY REFORM in 1831 and 1832. 16mo. 1s.

A DICTIONARY of the ENGLISH LANGUAGE. By R. G. LATHAM, M.A. M.D. F.R.S. Founded on the Dictionary of Dr. S. JOHNSON, as edited by the Rev. H. J. TODD, with numerous Emendations and Additions. Publishing in 36 Parts, price 3s. 6d. each, to form 2 vols. 4to.

THESAURUS of ENGLISH WORDS and PHRASES, classified and arranged so as to facilitate the Expression of Ideas, and assist in Literary Composition. By P. M. ROGET, M.D. 18th Edition. Crown 8vo. 10s. 6d.

LECTURES on the SCIENCE of LANGUAGE, delivered at the Royal Institution. By MAX MÜLLER, M.A. Taylorian Professor in the University of Oxford. FIRST SERIES, Fourth Edition, 12s. SECOND SERIES, 18s.

CHAPTERS on LANGUAGE. By FREDERIC W. FARRAR, M.A. late Fellow of Trin. Coll. Cambridge, Author of 'The Origin of Language,' &c. Crown 8vo. 8s. 6d.

The DEBATER; a Series of Complete Debates, Outlines of Debates, and Questions for Discussion. By F. ROWTON. Fcp. 6s.

A COURSE of ENGLISH READING, adapted to every taste and capacity; or, How and What to Read. By the Rev. J. PYCROFT, B.A. Fourth Edition. Fcp. 5s.

MANUAL of ENGLISH LITERATURE, Historical and Critical: with a Chapter on English Metres. By THOMAS ARNOLD, B.A. Post 8vo. 10s. 6d.

SOUTHEY'S DOCTOR, complete in One Volume. Edited by the Rev. J. W. WARTER, B.D. Square crown 8vo. 12s. 6d.

HISTORICAL and CRITICAL COMMENTARY on the OLD TESTAMENT; with a New Translation. By M. M. KALISCH, Ph.D. VOL. I. *Genesis*, 8vo. 18s. or adapted for the General Reader, 12s. VOL. II. *Exodus*, 15s. or adapted for the General Reader, 12s.

A Hebrew Grammar, with Exercises. By the same. PART I. *Outlines with Exercises*, 8vo. 12s. 6d. KEY, 5s. PART II. *Exceptional Forms and Constructions*, 12s. 6d.

A LATIN-ENGLISH DICTIONARY. By J. T. WHITE, M.A. of Corpus Christi College, and J. E. RIDDLE, M.A. of St. Edmund Hall, Oxford. Imperial 8vo. pp. 2,128, price 42s. cloth.

A New Latin-English Dictionary, abridged from the larger work of *White* and *Riddle* (as above), by J. T. WHITE, M.A. Joint-Author. Medium 8vo. pp. 1,048, price 18s. cloth.

The Junior Scholar's Latin-English Dictionary, abridged from the larger works of *White* and *Riddle* (as above), by J. T. White, M.A. surviving Joint-Author. Square 12mo. pp. 662, price 7s. 6d. cloth.

An **ENGLISH-GREEK LEXICON**, containing all the Greek Words used by Writers of good authority. By C. D. YONGE, B.A. Fifth Edition. 4to. 21s.

Mr. **YONGE'S NEW LEXICON**, English and Greek, abridged from his larger work (as above). Revised Edition. Square 12mo. 8s. 6d.

A **GREEK-ENGLISH LEXICON.** Compiled by H. G. LIDDELL, D.D. Dean of Christ Church, and R. SCOTT, D.D. Master of Balliol. Fifth Edition. Crown 4to. 31s. 6d.

A **Lexicon, Greek and English**, abridged from LIDDELL and SCOTT'S *Greek-English Lexicon*. Eleventh Edition. Square 12mo. 7s. 6d.

A **SANSKRIT-ENGLISH DICTIONARY**, the Sanskrit words printed both in the original Devanagari and in Roman letters; with References to the Best Editions of Sanskrit Authors, and with Etymologies and Comparisons of Cognate Words chiefly in Greek, Latin, Gothic, and Anglo-Saxon. Compiled by T. BENFEY, Prof. in the Univ. of Göttingen. 8vo. 52s. 6d.

A **PRACTICAL DICTIONARY of the FRENCH and ENGLISH LANGUAGES.** By L. CONTANSEAU. Tenth Edition. Post 8vo. 10s. 6d.

Contanseau's Pocket Dictionary, French and English, abridged from the above by the Author. Third Edition. 18mo. 5s.

NEW PRACTICAL DICTIONARY of the GERMAN LANGUAGE; German-English and English-German. By the Rev. W. L. BLACKLEY, M.A. and Dr. CARL MARTIN FRIEDLANDER. Post 8vo. [*Nearly ready.*

Miscellaneous Works and *Popular Metaphysics.*

RECREATIONS of a COUNTRY PARSON. By A. K. H. B. FIRST SERIES, with 41 Woodcut Illustrations from Designs by R. T. Pritchett. Crown 8vo. 12s. 6d.

Recreations of a Country Parson. SECOND SERIES. Cr. 8vo. 3s. 6d.

The Common-place Philosopher in Town and Country. By the same Author. Crown 8vo. 3s. 6d.

Leisure Hours in Town; Essays Consolatory, Æsthetical, Moral, Social, and Domestic. By the same Author. Crown 8vo. 3s. 6d.

The Autumn Holidays of a Country Parson; Essays contributed to *Fraser's Magazine* and to *Good Words.* By the same. Crown 8vo. 3s. 6d.

The Graver Thoughts of a Country Parson. SECOND SERIES. By the same Author. Crown 8vo. 3s. 6d.

Critical Essays of a Country Parson. Selected from Essays contributed to *Fraser's Magazine*. By the same Author. Post 8vo. 9s.

A **CAMPAIGNER AT HOME.** By SHIRLEY, Author of 'Thalatta' and 'Nugæ Criticæ.' Post 8vo. with Vignette, 7s. 6d.

STUDIES in PARLIAMENT. A Series of Sketches of Leading Politicians. By R. H. HUTTON. [Reprinted from the 'Pall Mall Gazette.'] Crown 8vo. 4s. 6d.

LORD MACAULAY'S MISCELLANEOUS WRITINGS.
 LIBRARY EDITION. 2 vols. 8vo. Portrait, 21s.
 PEOPLE'S EDITION. 1 vol. crown 8vo. 4s. 6d.

The REV. SYDNEY SMITH'S MISCELLANEOUS WORKS; including his Contributions to the *Edinburgh Review*.
 LIBRARY EDITION, 3 vols. 8vo. 36s.
 TRAVELLER'S EDITION, in 1 vol. 21s.
 CABINET EDITION, 3 vols. fcp. 21s.
 PEOPLE'S EDITION, 2 vols. crown 8vo. 8s.

Elementary Sketches of Moral Philosophy, delivered at the Royal Institution. By the same Author. Fcp. 7s.

The Wit and Wisdom of the Rev. Sydney Smith: a Selection of the most memorable Passages in his Writings and Conversation. 16mo. 5s.

EPIGRAMS, Ancient and Modern; Humorous, Witty, Satirical, Moral, and Panegyrical. Edited by Rev. JOHN BOOTH, B.A. Cambridge. Second Edition, revised and enlarged. Fcp. 7s. 6d.

From MATTER to SPIRIT: the Result of Ten Years' Experience in Spirit Manifestations. By SOPHIA E. DE MORGAN. With a PREFACE by Professor DE MORGAN. Post 8vo. 8s. 6d.

ESSAYS selected from CONTRIBUTIONS to the *Edinburgh Review*. By HENRY ROGERS. Second Edition. 3 vols. fcp. 21s.

The Eclipse of Faith; or, a Visit to a Religious Sceptic. By the same Author. Eleventh Edition. Fcp. 5s.

Defence of the Eclipse of Faith, by its Author; a rejoinder to Dr. Newman's *Reply*. Third Edition. Fcp. 3s. 6d.

Selections from the Correspondence of R. E. H. Greyson. By the same Author. Third Edition. Crown 8vo. 7s. 6d.

Fulleriana, or the Wisdom and Wit of THOMAS FULLER, with Essay on his Life and Genius. By the same Author. 16mo. 2s. 6d.

An ESSAY on HUMAN NATURE; showing the Necessity of a Divine Revelation for the Perfect Development of Man's Capacities. By HENRY S. BOASE, M.D. F.R.S. and G.S.

The PHILOSOPHY of NATURE; a Systematic Treatise on the Causes and Laws of Natural Phænomena. By the same Author. 8vo. 12s.

An INTRODUCTION to MENTAL PHILOSOPHY, on the Inductive Method. By. J. D. MORELL, M.A. LL.D. 8vo. 12s.

Elements of Psychology, containing the Analysis of the Intellectual Powers. By the same Author. Post 8vo. 7s. 6d.

The **SECRET of HEGEL**: being the Hegelian System in Origin, Principle, Form, and Matter. By James Hutchison Stirling. 2 vols. 8vo. 28s.

SIGHT and TOUCH: an Attempt to Disprove the Received (or Berkeleian) Theory of Vision. By Thomas K. Abbott, M.A. Fellow and Tutor of Trin. Coll. Dublin. 8vo. with 21 Woodcuts, 5s. 6d.

The **SENSES and the INTELLECT**. By Alexander Bain, M.A. Professor of Logic in the University of Aberdeen. Second Edition. 8vo. price 15s.

The **Emotions and the Will**, by the same Author; completing a Systematic Exposition of the Human Mind. 8vo. 15s.

On the **Study of Character**, including an Estimate of Phrenology. By the same Author. 8vo. 9s.

TIME and SPACE: a Metaphysical Essay. By Shadworth H. Hodgson. 8vo. pp. 588, price 16s.

The **WAY to REST**: Results from a Life-search after Religious Truth. By R. Vaughan, D.D.

HOURS WITH THE MYSTICS: a Contribution to the History of Religious Opinion. By Robert Alfred Vaughan, B.A. Second Edition. 2 vols. crown 8vo. 12s.

The **PHILOSOPHY of NECESSITY**; or Natural Law as applicable to Mental, Moral, and Social Science. By Charles Bray. Second Edition. 8vo. 9s.

The **Education of the Feelings and Affections**. By the same Author. Third Edition. 8vo. 3s. 6d.

CHRISTIANITY and COMMON SENSE. By Sir Willoughby Jones, Bart. M.A. Trin. Coll. Cantab. 8vo. 6s.

Astronomy, Meteorology, Popular Geography, &c.

OUTLINES of ASTRONOMY. By Sir J. F. W. Herschel, Bart. M.A. Eighth Edition, revised; with Plates and Woodcuts. 8vo. 18s.

ARAGO'S POPULAR ASTRONOMY. Translated by Admiral W. H. Smyth, F.R.S. and R. Grant, M.A. With 25 Plates and 358 Woodcuts. 2 vols. 8vo. £2 5s.

SATURN and its SYSTEM. By Richard A. Proctor, B.A. late Scholar of St John's Coll. Camb. and King's Coll. London. 8vo. with 14 Plates, 14s.

CELESTIAL OBJECTS for COMMON TELESCOPES. By the Rev. T. W. Webb, M.A. F.R.A.S. With Map of the Moon, and Woodcuts. 16mo. 7s.

PHYSICAL GEOGRAPHY for SCHOOLS and GENERAL READERS. By M. F. Maury, LL.D. Fcp. with 2 Charts, 2s. 6d.

M'CULLOCH'S DICTIONARY, Geographical, Statistical, and Historical, of the various Countries, Places, and Principal Natural Objects in the World. Revised Edit. printed in a larger type, with Maps, and with the Statistical Information throughout brought up to the latest returns by F. MARTIN. 4 vols. 8vo. 21s. each. VOL. I. now ready.

A GENERAL DICTIONARY of GEOGRAPHY, Descriptive, Physical, Statistical, and Historical; forming a complete Gazetteer of the World. By A. KEITH JOHNSTON, F.R.S.E. 8vo. 31s. 6d.

A MANUAL of GEOGRAPHY, Physical, Industrial, and Political. By W. HUGHES, F.R.G.S. Professor of Geography in King's College, and in Queen's College, London. With 6 Maps. Fcp. 7s. 6d.

The Geography of British History; a Geographical Description of the British Islands at Successive Periods. By the same. With 6 Maps. Fcp. 8s. 6d.

Abridged Text-Book of British Geography. By the same. Fcp. 1s. 6d.

MAUNDER'S TREASURY of GEOGRAPHY, Physical, Historical, Descriptive, and Political. Edited by W. HUGHES, F.R.G.S. With 7 Maps and 16 Plates. Fcp. 10s. 6d.

Natural History and *Popular Science.*

The ELEMENTS of PHYSICS or NATURAL PHILOSOPHY. By NEIL ARNOTT, M.D. F.R.S. Physician Extraordinary to the Queen. Sixth Edition, rewritten and completed. 2 Parts, 8vo. 21s.

HEAT CONSIDERED as a MODE of MOTION. By Professor JOHN TYNDALL, LL.D. F.R.S. Second Edition. Crown 8vo. with Woodcuts, 12s. 6d.

VOLCANOS, the Character of their Phenomena, their Share in the Structure and Composition of the Surface of the Globe, &c. By G. POULETT SCROPE, M.P. F.R.S. Second Edition. 8vo. with Illustrations, 15s.

A TREATISE on ELECTRICITY, in Theory and Practice. By A. DE LA RIVE, Prof. in the Academy of Geneva. Translated by C. V. WALKER, F.R.S. 3 vols. 8vo. with Woodcuts, £3 13s.

The CORRELATION of PHYSICAL FORCES. By W. R. GROVE, Q.C. V.P.R.S. Fourth Edition. 8vo. 7s. 6d.

MANUAL of GEOLOGY. By S. HAUGHTON, M.D. F.R.S. Fellow of Trin. Coll. and Prof. of Geol. in the Univ. of Dublin. Revised Edition, with 66 Woodcuts. Fcp. 6s.

A GUIDE to GEOLOGY. By J. PHILLIPS, M.A. Professor of Geology in the University of Oxford. Fifth Edition, with Plates. Fcp. 4s.

A GLOSSARY of MINERALOGY. By H. W. BRISTOW, F.G.S. of the Geological Survey of Great Britain. With 486 Figures. Crown 8vo. 12s.

PHILLIPS'S ELEMENTARY INTRODUCTION to MINERALOGY, with extensive Alterations and Additions, by H. J. BROOKE, F.R.S. and W. H. MILLER, F.G.S. Post 8vo. with Woodcuts, 18s.

VAN DER HOEVEN'S HANDBOOK of ZOOLOGY. Translated from the Second Dutch Edition by the Rev. W. CLARK, M.D. F.R.S. 2 vols. 8vo. with 24 Plates of Figures, 60s.

The COMPARATIVE ANATOMY and PHYSIOLOGY of the VERTEbrate Animals. By RICHARD OWEN, F.R.S. D.C.L. 3 vols. 8vo. with upwards of 1,200 Woodcuts. VOLS. I. and II. price 21s. each, now ready.

HOMES WITHOUT HANDS: a Description of the Habitations of Animals, classed according to their Principle of Construction. By Rev. J. G. WOOD, M.A. F.L.S. With about 140 Vignettes on Wood (20 full size of page). Second Edition. 8vo. 21s.

MANUAL of CORALS and SEA JELLIES. By J. R. GREENE, B.A. Edited by the Rev. J. A. GALBRAITH, M.A. and the Rev. S. HAUGHTON, M.D. Fcp. with 39 Woodcuts, 5s.

Manual of Sponges and Animalculæ; with a General Introduction on the Principles of Zoology. By the same Author and Editors. Fcp. with 16 Woodcuts, 2s.

Manual of the Metalloids. By J. APJOHN, M.D. F.R.S. and the same Editors. Revised Edition. Fcp. with 38 Woodcuts, 7s. 6d.

The HARMONIES of NATURE and UNITY of CREATION. By Dr. GEORGE HARTWIG. 8vo. with numerous Illustrations.

The Sea and its Living Wonders. By the same Author. Second (English) Edition. 8vo. with many Illustrations. 18s.

The Tropical World. By the same Author. With 8 Chromoxylographs and 172 Woodcuts. 8vo. 21s.

SKETCHES of the NATURAL HISTORY of CEYLON. By Sir J. EMERSON TENNENT, K.C.S. LL.D. With 82 Wood Engravings. Post 8vo. price 12s. 6d.

Ceylon. By the same Author. Fifth Edition; with Maps, &c. and 90 Wood Engravings. 2 vols. 8vo. £2 10s

A FAMILIAR HISTORY of BIRDS. By E. STANLEY, D.D. F.R.S. late Lord Bishop of Norwich. Seventh Edition, with Woodcuts. Fcp. 3s. 6d.

MARVELS and MYSTERIES of INSTINCT; or, Curiosities of Animal Life. By G. GARRATT. Third Edition. Fcp. 7s.

HOME WALKS and HOLIDAY RAMBLES. By the Rev. C. A. JOHNS, B.A. F.L.S. Fcp. 8vo. with 10 Illustrations, 6s.

KIRBY and SPENCE'S INTRODUCTION to ENTOMOLOGY, or Elements of the Natural History of Insects. Seventh Edition. Crown 8vo. price 5s.

MAUNDER'S TREASURY of NATURAL HISTORY, or Popular Dictionary of Zoology. Revised and corrected by T. S. COBBOLD, M.D. Fcp. with 900 Woodcuts, 10s.

The TREASURY of BOTANY, or Popular Dictionary of the Vegetable Kingdom; with which is incorporated a Glossary of Botanical Terms. Edited by J. LINDLEY, F.R.S. and T. MOORE, F.L.S. assisted by eminent Contributors. Pp. 1,274, with 274 Woodcuts and 20 Steel Plates. 2 Parts, fcp. 20s.

The **ELEMENTS** of **BOTANY** for **FAMILIES** and **SCHOOLS**.
Tenth Edition, revised by Thomas Moore, F.L.S. Fcp. with 154 Woodcuts, 2s. 6d.

The **ROSE AMATEUR'S GUIDE.** By Thomas Rivers. New Edition. Fcp. 4s.

The **BRITISH FLORA**; comprising the Phænogamous or Flowering Plants and the Ferns. By Sir W. J. Hooker, K.H. and G. A. Walker-Arnott, LL.D. 12mo. with 12 Plates, 14s. or coloured, 21s.

BRYOLOGIA BRITANNICA; containing the Mosses of Great Britain and Ireland, arranged and described. By W. Wilson. 8vo. with 61 Plates 42s. or coloured, £4 4s.

The **INDOOR GARDENER.** By Miss Maling. Fcp. with Frontispiece, printed in Colours, 5s.

LOUDON'S ENCYCLOPÆDIA of PLANTS; comprising the Specific Character, Description, Culture, History, &c. of all the Plants found in Great Britain. With upwards of 12,000 Woodcuts. 8vo. £3 13s. 6d.

Loudon's Encyclopædia of Trees and Shrubs; containing the Hardy Trees and Shrubs of Great Britain scientifically and popularly described. With 2,000 Woodcuts. 8vo. 50s.

MAUNDER'S SCIENTIFIC and LITERARY TREASURY; a Popular Encyclopædia of Science, Literature, and Art. Fcp. 10s.

A DICTIONARY of SCIENCE, LITERATURE, and ART. Fourth Edition, re-edited by W. T. Brande (the Author), and George W. Cox. M.A. assisted by gentlemen of eminent Scientific and Literary Acquirements. 3 vols. medium 8vo. price 63s. cloth.

ESSAYS on SCIENTIFIC and other SUBJECTS, contributed to Reviews. By Sir H. Holland, Bart. M.D. Second Edition. 8vo. 14s.

ESSAYS from the EDINBURGH and QUARTERLY REVIEWS; with Addresses and other Pieces. By Sir J. F. W. Herschel, Bart. M.A. 8vo. 18s.

Chemistry, Medicine, Surgery, and the *Allied Sciences.*

A DICTIONARY of CHEMISTRY and the Allied Branches of other Sciences; founded on that of the late Dr. Ure. By Henry Watts, F.C.S. assisted by eminent Contributors. 5 vols. medium 8vo. in course of publication in Parts. Vol. I. 31s. 6d. Vol. II. 26s. Vol. III. 31s. 6d. are now ready.

HANDBOOK of CHEMICAL ANALYSIS. Adapted to the Unitary System of Notation. By F. T. Conington, M.A. F.C.S. Post 8vo. 7s. 6d.—Tables of Qualitative Analysis adapted to the same, 2s. 6d.

A HANDBOOK of VOLUMETRICAL ANALYSIS. By Robert H. Scott, M.A. T.C.D. Post 8vo. 4s. 6d.

ELEMENTS of CHEMISTRY, Theoretical and Practical. By WILLIAM A. MILLER, M.D. LL.D. F.R.S. F.G.S. Professor of Chemistry, King's College, London. 3 vols. 8vo. £2 13s. PART I. CHEMICAL PHYSICS. Third Edition, 12s. PART II. INORGANIC CHEMISTRY, 21s. PART III. ORGANIC CHEMISTRY, Second Edition, 20s.

A MANUAL of CHEMISTRY, Descriptive and Theoretical. By WILLIAM ODLING, M.B. F.R.S. PART I. 8vo. 9s.

A Course of Practical Chemistry, for the use of Medical Students. By the same Author. Second Edition, with 70 new Woodcuts. Crown 8vo. price 7s. 6d.

Lectures on Animal Chemistry, delivered at the Royal College of Physicians in 1865. By the same Author. Crown 8vo. 4s. 6d.

The DIAGNOSIS and TREATMENT of the DISEASES of WOMEN; including the Diagnosis of Pregnancy. By GRAILY HEWITT, M.D. 8vo. 16s.

LECTURES on the DISEASES of INFANCY and CHILDHOOD. By CHARLES WEST, M.D. &c. Fifth Edition, revised and enlarged. 8vo. 16s.

EXPOSITION of the SIGNS and SYMPTOMS of PREGNANCY: with other Papers on subjects connected with Midwifery. By W. F. MONTGOMERY, M.A. M.D. M.R.I.A. 8vo. with Illustrations, 25s.

A SYSTEM of SURGERY, Theoretical and Practical. In Treatises by Various Authors. Edited by T. HOLMES, M.A. Cantab. Assistant-Surgeon to St. George's Hospital. 4 vols. 8vo. £4 13s.

Vol. I. General Pathology. 21s.

Vol. II. Local Injuries: Gunshot Wounds, Injuries of the Head, Back, Face, Neck, Chest, Abdomen, Pelvis, of the Upper and Lower Extremities, and Diseases of the Eye. 21s.

Vol. III. Operative Surgery. Diseases of the Organs of Circulation, Locomotion, &c. 21s.

Vol. IV. Diseases of the Organs of Digestion, of the Genito-Urinary System, and of the Breast, Thyroid Gland, and Skin; with APPENDIX and GENERAL INDEX. 30s.

LECTURES on the PRINCIPLES and PRACTICE of PHYSIC. By THOMAS WATSON, M.D. Physician-Extraordinary to the Queen. Fourth Edition. 2 vols. 8vo. 34s.

LECTURES on SURGICAL PATHOLOGY. By J. PAGET, F.R.S. Surgeon-Extraordinary to the Queen. Edited by W. TURNER, M.B. 8vo. with 117 Woodcuts, 21s.

A TREATISE on the CONTINUED FEVERS of GREAT BRITAIN. By C. MURCHISON, M.D. Senior Physician to the London Fever Hospital. 8vo. with coloured Plates, 18s.

ANATOMY, DESCRIPTIVE and SURGICAL. By HENRY GRAY, F.R.S. With 410 Wood Engravings from Dissections. Third Edition, by T. HOLMES, M.A. Cantab. Royal 8vo. 28s.

The CYCLOPÆDIA of ANATOMY and PHYSIOLOGY. Edited by the late R. B. TODD, M.D. F.R.S. Assisted by nearly all the most eminent cultivators of Physiological Science of the present age. 5 vols. 8vo. with 2,853 Woodcuts, £6 6s.

PHYSIOLOGICAL ANATOMY and PHYSIOLOGY of MAN. By the late R. B. TODD, M.D. F.R.S. and W. BOWMAN, F.R.S. of King's College. With numerous Illustrations. Vol. II. 8vo. 25s.

A DICTIONARY of PRACTICAL MEDICINE. By J. COPLAND, M.D. F.R.S. Abridged from the larger work by the Author, assisted by J. C. COPLAND, M.R.C.S. and throughout brought down to the present State of Medical Science. Pp. 1,560 in 8vo. price 36s.

Dr. Copland's Dictionary of Practical Medicine (the larger work). 3 vols. 8vo. £5 11s.

The WORKS of SIR B. C. BRODIE, Bart. collected and arranged by CHARLES HAWKINS, F.R.C.S.E. 3 vols. 8vo. with Medallion and Facsimile, 48s.

Autobiography of Sir B. C. Brodie, Bart. Printed from the Author's materials left in MS. Second Edition. Fcp. 4s. 6d.

The TOXICOLOGIST'S GUIDE: a New Manual on Poisons, giving the Best Methods to be pursued for the Detection of Poisons (post-mortem or otherwise). By JOHN HORSLEY, F.C.S. Analytical Chemist.

A MANUAL of MATERIA MEDICA and THERAPEUTICS, abridged from Dr. PEREIRA's *Elements* by F. J. FARRE, M.D. assisted by R. BENTLEY, M.R.C.S. and by R. WARINGTON, F.R.S. 8vo. with 90 Woodcuts, 21s.

Dr. Pereira's Elements of Materia Medica and Therapeutics. Third Edition. By A. S. TAYLOR, M.D. and G. O. REES, M.D. 3 vols. 8vo. with Woodcuts, £3 15s.

THOMSON'S CONSPECTUS of the BRITISH PHARMACOPŒIA. Twenty-fourth Edition, corrected and made conformable throughout to the New Pharmacopœia of the General Council of Medical Education. By E. LLOYD BIRKETT, M.D. 18mo. 5s. 6d.

MANUAL of the DOMESTIC PRACTICE of MEDICINE. By W. B. KESTEVEN, F.R.C.S.E. Second Edition, revised, with Additions. Fcp. 5s.

The RESTORATION of HEALTH; or, the Application of the Laws of Hygiene to the Recovery of Health: a Manual for the Invalid, and a Guide in the Sick Room. By W. STRANGE, M.D. Fcp. 6s.

SEA-AIR and SEA-BATHING for CHILDREN and INVALIDS By the same Author. Fcp. boards, 3s.

MANUAL for the CLASSIFICATION, TRAINING, and EDUCATION of the Feeble-Minded, Imbecile, and Idiotic. By P. MARTIN DUNCAN, M.B. and WILLIAM MILLARD. Crown 8vo. 5s.

The Fine Arts, and *Illustrated Editions.*

The NEW TESTAMENT, illustrated with Wood Engravings after the Early Masters, chiefly of the Italian School. Crown 4to. 63s. cloth, gilt top; or £5 5s. elegantly bound in morocco.

LYRA GERMANICA; Hymns for the Sundays and Chief Festivals of the Christian Year. Translated by CATHERINE WINKWORTH; 125 Illustrations on Wood drawn by J. LEIGHTON, F.S.A. Fcp. 4to. 21s.

The **LIFE of MAN SYMBOLISED by the MONTHS of the YEAR** in their Seasons and Phases; with Passages selected from Ancient and Modern Authors. By RICHARD PIGOT. Accompanied by a Series of 25 full-page Illustrations and numerous Marginal Devices, Decorative Initial Letters, and Tailpieces, engraved on Wood from Original Designs by JOHN LEIGHTON, F.S.A. 4to. 42s.

CATS' and FARLIE'S MORAL EMBLEMS; with Aphorisms, Adages, and Proverbs of all Nations: comprising 121 Illustrations on Wood by J. LEIGHTON, F.S.A. with an appropriate Text by R. PIGOT. Imperial 8vo. 31s. 6d.

SHAKSPEARE'S SENTIMENTS and SIMILES, printed in Black and Gold, and Illuminated in the Missal Style by HENRY NOEL HUMPHREYS. In massive covers, containing the Medallion and Cypher of Shakspeare. Square post 8vo. 21s.

The **HISTORY of OUR LORD**, as exemplified in Works of Art. Being the fourth and concluding series of 'Sacred and Legendary Art.' By Mrs. JAMESON and Lady EASTLAKE. Second Edition, with 13 Etchings and 281 Woodcuts. 2 vols. square crown 8vo. 42s.

In the same Series, by Mrs. JAMESON.

Legends of the Saints and Martyrs. Fourth Edition, with 19 Etchings and 187 Woodcuts. 2 vols. 31s. 6d.

Legends of the Monastic Orders. Third Edition, with 11 Etchings and 88 Woodcuts. 1 vol. 21s.

Legends of the Madonna. Third Edition, with 27 Etchings and 165 Woodcuts. 1 vol. 21s.

Arts, Manufactures, &c.

DRAWING from NATURE; a Series of Progressive Instructions in Sketching, from Elementary Studies to Finished Views, with Examples from Switzerland and the Pyrenees. By GEORGE BARNARD, Professor of Drawing at Rugby School. With 18 Lithographic Plates, and 108 Wood Engravings. Imp. 8vo. 25s.

ENCYCLOPÆDIA of ARCHITECTURE, Historical, Theoretical, and Practical. By JOSEPH GWILT. With more than 1,000 Woodcuts. 8vo. 42s.

TUSCAN SCULPTORS, their Lives, Works, and Times. With 45 Etchings and 28 Woodcuts from Original Drawings and Photographs. By CHARLES C. PERKINS. 2 vols. imperial 8vo. 63s.

The **GRAMMAR of HERALDRY**: containing a Description of all the Principal Charges used in Armory, the Signification of Heraldic Terms, and the Rules to be observed in Blazoning and Marshalling. By JOHN E. CUSSANS. Fcp. with 196 Woodcuts, 4s. 6d.

The **ENGINEER'S HANDBOOK**; explaining the Principles which should guide the young Engineer in the Construction of Machinery. By C. S. LOWNDES. Post 8vo. 5s.

The **ELEMENTS** of **MECHANISM**. By T. M. GOODEVE, M.A. Professor of Mechanics at the R. M. Acad. Woolwich. Second Edition, with 217 Woodcuts. Post 8vo. 6s. 6d.

URE'S DICTIONARY of ARTS, MANUFACTURES, and MINES. Re-written and enlarged by ROBERT HUNT, F.R.S. assisted by numerous gentlemen eminent in Science and the Arts. With 2,000 Woodcuts. 3 vols. 8vo. £4.

ENCYCLOPÆDIA of CIVIL ENGINEERING, Historical, Theoretical, and Practical. By E. CRESY, C.E. With above 3,000 Woodcuts. 8vo. 42s.

TREATISE on MILLS and MILLWORK. By W. FAIRBAIRN, C.E. Second Edition, with 18 Plates and 322 Woodcuts. 2 vols. 8vo. 32s.

Useful Information for Engineers. By the same Author. FIRST and SECOND SERIES, with many Plates and Woodcuts. 2 vols. crown 8vo. 10s. 6d. each.

The Application of Cast and Wrought Iron to Building Purposes. By the same Author. Third Edition, with 6 Plates and 118 Woodcuts. 8vo. 16s.

IRON SHIP BUILDING, its History and Progress, as comprised in a Series of Experimental Researches on the Laws of Strain; the Strengths, Forms, and other conditions of the Material; and an Inquiry into the Present and Prospective State of the Navy, including the Experimental Results on the Resisting Powers of Armour Plates and Shot at High Velocities. By the same Author. With 4 Plates and 130 Woodcuts. 8vo. 18s.

The PRACTICAL MECHANIC'S JOURNAL: an Illustrated Record of Mechanical and Engineering Science, and Epitome of Patent Inventions. 4to. price 1s. monthly.

The PRACTICAL DRAUGHTSMAN'S BOOK of INDUSTRIAL DESIGN. By W. JOHNSON, Assoc. Inst. C.E. With many hundred Illustrations. 4to. 28s. 6d.

The PATENTEE'S MANUAL. a Treatise on the Law and Practice of Letters Patent for the use of Patentees and Inventors. By J. and J. H. JOHNSON. Post 8vo. 7s. 6d.

The ARTISAN CLUB'S TREATISE on the STEAM ENGINE, in its various Applications to Mines, Mills, Steam Navigation, Railways and Agriculture. By J. BOURNE, C.E. Seventh Edition; with 37 Plates and 546 Woodcuts. 4to. 42s.

Catechism of the Steam Engine, in its various Applications to Mines, Mills, Steam Navigation, Railways, and Agriculture. By the same Author. With 199 Woodcuts. Fcp. 9s. The INTRODUCTION of 'Recent Improvements' may be had separately, with 110 Woodcuts, price 3s. 6d.

Handbook of the Steam Engine. By the same Author, forming a KEY to the Catechism of the Steam Engine, with 67 Woodcuts. Fcp. 9s.

A TREATISE on the SCREW PROPELLER, SCREW VESSELS, and Screw Engines, as adapted for purposes of Peace and War; illustrated by many Plates and Woodcuts. By the same Author. New and enlarged Edition, in course of publication in 24 Parts. Royal 4to. 2s. 6d. each.

The THEORY of WAR Illustrated by numerous Examples from History. By Lieut.-Col. P. L. MACDOUGALL. Third Edition, with 10 Plans. Post 8vo. 10s. 6d.

The **ART of PERFUMERY**; the History and Theory of Odours, and the Methods of Extracting the Aromas of Plants. By Dr. PIESSE, F.C.S. Third Edition, with 53 Woodcuts. Crown 8vo. 10s. 6d.

Chemical, Natural, and Physical Magic, for Juveniles during the Holidays. By the same Author. Third Edition, enlarged, with 38 Woodcuts. Fcp. 6s.

TALPA; or the Chronicles of a Clay Farm. By C. W. HOSKYNS, Esq. Sixth Edition, with 24 Woodcuts by G. CRUIKSHANK. 16mo. 5s. 6d.

LOUDON'S ENCYCLOPÆDIA of AGRICULTURE: comprising the Laying-out, Improvement, and Management of Landed Property, and the Cultivation and Economy of the Productions of Agriculture. With 1,100 Woodcuts. 8vo. 31s. 6d.

Loudon's Encylopædia of Gardening: comprising the Theory and Practice of Horticulture, Floriculture, Arboriculture, and Landscape Gardening. With 1,000 Woodcuts. 8vo. 31s. 6d.

Loudon's Encyclopædia of Cottage, Farm, and Villa Architecture and Furniture. With more than 2,000 Woodcuts. 8vo. 42s.

HISTORY of WINDSOR GREAT PARK and WINDSOR FOREST. By WILLIAM MENZIES, Resident Deputy Surveyor. With 2 Maps and 20 Photographs. Imp. folio, £8 8s.

BAYLDON'S ART of VALUING RENTS and TILLAGES, and Claims of Tenants upon Quitting Farms, both at Michaelmas and Lady-Day. Eighth Edition, revised by J. C. MORTON. 8vo. 10s. 6d.

Religious and *Moral Works.*

An EXPOSITION of the 39 ARTICLES, Historical and Doctrinal. By E. HAROLD BROWNE, D.D. Lord Bishop of Ely. Seventh Edit. 8vo. 16s.

The Pentateuch and the Elohistic Psalms, in Reply to Bishop Colenso. By the same. Second Edition. 8vo. 2s.

Examination Questions on Bishop Browne's Exposition of the Articles. By the Rev. J. GORLE, M.A. Fcp. 3s. 6d.

FIVE LECTURES on the CHARACTER of ST. PAUL; being the Hulsean Lectures for 1862. By the Rev. J. S. HOWSON, D.D. Second Edition. 8vo. 9s.

The LIFE and EPISTLES of ST. PAUL. By W. J. CONYBEARE, M.A. late Fellow of Trin. Coll. Cantab. and J. S. HOWSON, D.D. late Principal of Liverpool College.

LIBRARY EDITION, with all the Original Illustrations, Maps, Landscapes on Steel, Woodcuts, &c. 2 vols. 4to. 48s.

INTERMEDIATE EDITION, with a Selection of Maps, Plates, and Woodcuts. 2 vols. square crown 8vo. 31s. 6d.

PEOPLE'S EDITION, revised and condensed, with 46 Illustrations and Maps. 2 vols. crown 8vo. 12s.

The VOYAGE and SHIPWRECK of ST. PAUL; with Dissertations on the Ships and Navigation of the Ancients. By JAMES SMITH, F.R.S. Crown 8vo. Charts, 8s. 6d.

FASTI SACRI, or a Key to the Chronology of the New Testament; comprising an Historical Harmony of the Four Gospels, and Chronological Tables generally from B.C. 70 to A.D. 70; with a Preliminary Dissertation on the Chronology of the New Testament, and other Aids to the elucidation of the subject. By THOMAS LEWIN, M.A. F.S.A. Imperial 8vo. 42s.

A CRITICAL and GRAMMATICAL COMMENTARY on ST. PAUL'S Epistles. By C. J. ELLICOTT, D.D. Lord Bishop of Gloucester and Bristol. 8vo.

Galatians, Third Edition, 8s. 6d.

Ephesians, Third Edition, 8s. 6d.

Pastoral Epistles, Third Edition, 10s. 6d.

Philippians, Colossians, and Philemon, Third Edition, 10s. 6d.

Thessalonians, Second Edition, 7s. 6d.

Historical Lectures on the Life of our Lord Jesus Christ: being the Hulsean Lectures for 1859. By the same Author. Fourth Edition. 8vo. price 10s. 6d.

The Destiny of the Creature; and other Sermons preached before the University of Cambridge. By the same. Fourth Edition. Post 8vo. 5s.

The Broad and the Narrow Way; Two Sermons preached before the University of Cambridge. By the same. Crown 8vo. 2s.

Rev. T. H. HORNE'S INTRODUCTION to the CRITICAL STUDY and Knowledge of the Holy Scriptures. Eleventh Edition, corrected and extended under careful Editorial revision. With 4 Maps and 22 Woodcuts and Facsimiles. 4 vols. 8vo. £3 13s. 6d.

Rev. T. H. Horne's Compendious Introduction to the Study of the Bible, being an Analysis of the larger work by the same Author. Re-edited by the Rev. JOHN AYRE, M.A. With Maps. &c. Post 8vo. 9s.

The TREASURY of BIBLE KNOWLEDGE; being a Dictionary of the Books, Persons, Places, Events, and other matters of which mention is made in Holy Scripture; intended to establish its Authority and illustrate its Contents. By Rev. J. AYRE, M.A. With Maps, 16 Plates, and numerous Woodcuts. Fcp. 10s. 6d.

The GREEK TESTAMENT; with Notes, Grammatical and Exegetical. By the Rev. W. WEBSTER, M.A. and the Rev. W. F. WILKINSON, M.A. 2 vols. 8vo. £2 4s.

    VOL. I. the Gospels and Acts, 20s.

    VOL. II. the Epistles and Apocalypse, 24s.

EVERY-DAY SCRIPTURE DIFFICULTIES explained and illustrated. By J. E. PRESCOTT, M.A. VOL. I. *Matthew* and *Mark*; VOL. II. *Luke* and *John*. 2 vols. 8vo. 9s. each.

The PENTATEUCH and BOOK of JOSHUA CRITICALLY EXAMINED. By the Right Rev. J. W. COLENSO, D.D. Lord Bishop of Natal. People's Edition, in 1 vol. crown 8vo. 6s. or in 5 Parts, 1s. each.

The PENTATEUCH and BOOK of JOSHUA CRITICALLY EXAMINED. By Prof. A. KUENEN, of Leyden. Translated from the Dutch, and edited with Notes, by J. W. COLENSO, D.D. Bishop of Natal. 8vo. 8s. 6d.

The CHURCH and the WORLD: Essays on Questions of the Day. By Various Writers. Edited by the Rev. ORBY SHIPLEY, M.A. 8vo.

The **FORMATION of CHRISTENDOM.** Part I. By T. W. Allies, 8vo. 12s.

CHRISTENDOM'S DIVISIONS: a Philosophical Sketch of the Divisions of the Christian Family in East and West. By Edmund S. Ffoulkes, formerly Fellow and Tutor of Jesus Coll. Oxford. Post 8vo. 7s. 6d.

Christendom's Divisions, Part II. Greeks and Latins, being a History of their Dissensions and Overtures for Peace down to the Reformation. By the same Author. [*Nearly ready.*

The **LIFE of CHRIST:** an Eclectic Gospel, from the Old and New Testaments, arranged on a New Principle, with Analytical Tables, &c. By Charles De La Pryme, M.A. Trin. Coll. Camb. Revised Edition. 8vo. 5s.

The **HIDDEN WISDOM of CHRIST** and the **KEY of KNOWLEDGE;** or, History of the Apocrypha. By Ernest de Bunsen. 2 vols. 8vo. 28s.

ESSAYS on RELIGION and LITERATURE. Edited by the Most Rev. Archbishop Manning. 8vo. 10s. 6d.

The **TEMPORAL MISSION of the HOLY GHOST;** or, Reason and Revelation. By the Most Rev. Archbishop Manning. Second Edition. Crown 8vo. 8s. 6d.

ESSAYS and REVIEWS. By the Rev. W. Temple, D.D. the Rev. R. Williams, B.D. the Rev. B. Powell, M.A. the Rev. H. B. Wilson, B.D. C. W. Goodwin, M.A. the Rev. M. Pattison, B.D. and the Rev. B. Jowett, M.A. Twelfth Edition. Fcp. 8vo. 5s.

MOSHEIM'S ECCLESIASTICAL HISTORY. Murdock and Soames's Translation and Notes, re-edited by the Rev. W. Stubbs, M.A. 3 vols. 8vo. 45s.

BISHOP JEREMY TAYLOR'S ENTIRE WORKS: With Life by Bishop Heber. Revised and corrected by the Rev. C. P. Eden, 10 vols. price £5 5s.

PASSING THOUGHTS on RELIGION. By the Author of 'Amy Herbert.' New Edition. Fcp. 8vo. 5s.

Thoughts for the Holy Week, for Young Persons. By the same Author. Third Edition. Fcp. 8vo. 2s.

Night Lessons from Scripture. By the same Author. Second Edition. 32mo. 3s.

Self-Examination before Confirmation. By the same Author. 32mo. price 1s. 6d.

Readings for a Month Preparatory to Confirmation, from Writers of the Early and English Church. By the same. Fcp. 4s.

Readings for Every Day in Lent, compiled from the Writings of Bishop Jeremy Taylor. By the same. Fcp. 5s.

Preparation for the Holy Communion; the Devotions chiefly from the works of Jeremy Taylor. By the same. 32mo. 3s.

MORNING CLOUDS. Second Edition. Fcp. 5s.

PRINCIPLES of EDUCATION Drawn from Nature and Revelation, and applied to Female Education in the Upper Classes. By the same. 2 vols. fcp. 12s. 6d.

The **WIFE'S MANUAL**; or, Prayers, Thoughts, and Songs on Several Occasions of a Matron's Life. By the Rev. W. CALVERT, M.A. Crown 8vo. price 10s. 6d.

SPIRITUAL SONGS for the **SUNDAYS** and **HOLIDAYS** throughout the Year. By J. S. B. MONSELL, LL.D. Vicar of Egham. Fourth Edition. Fcp. 4s. 6d.

The **Beatitudes**: Abasement before God ; Sorrow for Sin ; Meekness of Spirit ; Desire for Holiness ; Gentleness ; Purity of Heart ; the Peacemakers ; Sufferings for Christ. By the same. Second Edition, fcp. 3s. 6d.

LYRA DOMESTICA; Christian Songs for Domestic Edification. Translated from the *Psaltery and Harp* of C. J. P. SPITTA, and from other sources, by RICHARD MASSIE. FIRST and SECOND SERIES, fcp. 4s. 6d. each.

LYRA SACRA; Hymns, Ancient and Modern, Odes and Fragments of Sacred Poetry. Edited by the Rev. B. W. SAVILE, M.A. Third Edition, enlarged and improved. Fcp. 5s.

LYRA GERMANICA, translated from the German by Miss C. WINKWORTH. FIRST SERIES, Hymns for the Sundays and Chief Festivals ; SECOND SERIES, the Christian Life. Fcp. 5s. each SERIES.

Hymns from **Lyra Germanica**, 18mo. 1s.

LYRA EUCHARISTICA; Hymns and Verses on the Holy Communion, Ancient and Modern ; with other Poems. Edited by the Rev. ORBY SHIPLEY, M.A. Second Edition. Fcp. 7s. 6d.

Lyra Messianica; Hymns and Verses on the Life of Christ, Ancient and Modern ; with other Poems. By the same Editor. Second Edition, altered and enlarged. Fcp. 7s. 6d.

Lyra Mystica; Hymns and Verses on Sacred Subjects, Ancient and Modern. By the same Editor. Fcp. 7s. 6d.

The **CHORALE BOOK** for **ENGLAND**; a complete Hymn-Book in accordance with the Services and Festivals of the Church of England : the Hymns translated by Miss C. WINKWORTH ; the tunes arranged by Prof. W. S. BENNETT and OTTO GOLDSCHMIDT. Fcp. 4to. 12s. 6d.

Congregational Edition. Fcp. 2s.

The **CATHOLIC DOCTRINE** of the **ATONEMENT**: an Historical Inquiry into its Development in the Church ; with an Introduction on the Principle of Theological Developments. By H. N. OXENHAM, M.A. formerly Scholar of Balliol College, Oxford. 8vo. 8s. 6d.

FROM SUNDAY TO SUNDAY: an attempt to consider familiarly the Weekday Life and Labours of a Country Clergyman. By R. GEE, M.A Vicar of Abbott's Langley and Rural Dean. Fcp. 5s.

FIRST SUNDAYS at CHURCH; or, Familiar Conversations on the Morning and Evening Services of the Church of England. By J. E. RIDDLE, M.A. Fcp. 2s. 6d.

The **JUDGMENT of CONSCIENCE**, and other Sermons. By RICHARD WHATELY, D.D. late Archbishop of Dublin. Crown 8vo. 4s. 6d.

PALEY'S MORAL PHILOSOPHY, with Annotations. By RICHARD WHATELY, D.D. late Archbishop of Dublin. 8vo. 7s.

Travels, Voyages, &c.

OUTLINE SKETCHES of the HIGH ALPS of DAUPHINÉ. By T. G. BONNEY, M.A. F.G.S. M.A.C. Fellow of St. John's Coll. Camb. With 13 Plates and a Coloured Map. Post 4to. 16s.

ICE-CAVES of FRANCE and SWITZERLAND; a Narrative of Subterranean Exploration. By the Rev. G. F. BROWNE, M.A. Fellow and Assistant-Tutor of St. Catherine's Coll. Cambridge, M.A.C. With 11 Illustrations on Wood. Square crown 8vo. 12s. 6d.

VILLAGE LIFE in SWITZERLAND. By SOPHIA D. DELMARD. Post 8vo. 9s. 6d.

HOW WE SPENT the SUMMER; or, a Voyage en Zigzag in Switzerland and Tyrol with some Members of the ALPINE CLUB. From the Sketch-Book of one of the Party. Third Edition, re-drawn. In oblong 4to. with about 300 Illustrations, 15s.

BEATEN TRACKS; or, Pen and Pencil Sketches in Italy. By the Authoress of 'A Voyage en Zigzag.' With 42 Plates, containing about 200 Sketches from Drawings made on the Spot. 8vo. 16s.

MAP of the CHAIN of MONT BLANC, from an actual Survey in 1863—1864. By A. ADAMS-REILLY, F.R.G.S. M.A.C. Published under the Authority of the Alpine Club. In Chromolithography on extra stout drawing-paper 28in. × 17in. price 10s. or mounted on canvas in a folding case, 12s. 6d.

TRANSYLVANIA, its PRODUCTS and its PEOPLE. By CHARLES BONER. With 5 Maps and 43 Illustrations on Wood and in Chromolithography. 8vo. 21s.

EXPLORATIONS in SOUTH WEST AFRICA, from Walvisch Bay to Lake Ngami and the Victoria Falls. By THOMAS BAINES, F.R.G.S. 8vo. with Map and Illustrations, 21s.

VANCOUVER ISLAND and BRITISH COLUMBIA; their History, Resources, and Prospects. By MATTHEW MACFIE, F.R.G.S. With Maps and Illustrations. 8vo. 18s.

HISTORY of DISCOVERY in our AUSTRALASIAN COLONIES, Australia, Tasmania, and New Zealand, from the Earliest Date to the Present Day. By WILLIAM HOWITT. With 3 Maps of the Recent Explorations from Official Sources. 2 vols. 8vo. 28s.

The CAPITAL of the TYCOON; a Narrative of a Three Years' Residence in Japan. By Sir RUTHERFORD ALCOCK, K.C.B. 2 vols. 8vo. with numerous Illustrations, 42s.

LAST WINTER in ROME. By C. R. WELD. With Portrait and Engravings on Wood. Post 8vo. 14s.

AUTUMN RAMBLES in NORTH AFRICA. By JOHN ORMSBY, of the Middle Temple. With 16 Illustrations. Post 8vo. 8s. 6d.

The DOLOMITE MOUNTAINS. Excursions through Tyrol, Carinthia, Carniola, and Friuli in 1861, 1862, and 1863. By J. GILBERT and G. C. CHURCHILL, F.R.G.S. With numerous Illustrations. Square crown 8vo. 21s.

A SUMMER TOUR in the GRISONS and ITALIAN VALLEYS of the Bernina. By Mrs. HENRY FRESHFIELD. With 2 Coloured Maps and 4 Views. Post 8vo. 10s. 6d.

Alpine Byeways; or, Light Leaves gathered in 1859 and 1860. By the same Authoress. Post 8vo. with Illustrations, 10s. 6d.

A LADY'S TOUR ROUND MONTE ROSA; including Visits to the Italian Valleys. With Map and Illustrations. Post 8vo. 14s.

GUIDE to the PYRENEES, for the use of Mountaineers. By CHARLES PACKE. With Maps, &c. and Appendix. Fcp. 6s.

The ALPINE GUIDE. By JOHN BALL, M.R.I.A. late President of the Alpine Club. Post 8vo. with Maps and other Illustrations.

Guide to the Eastern Alps, *nearly ready*.

Guide to the Western Alps, including Mont Blanc, Monte Rosa, Zermatt, &c. 7s. 6d.

Guide to the Oberland and all Switzerland, excepting the Neighbourhood of Monte Rosa and the Great St. Bernard; with Lombardy and the adjoining portion of Tyrol. 7s. 6d.

A GUIDE to SPAIN. By H. O'SHEA. Post 8vo. with Travelling Map, 15s.

CHRISTOPHER COLUMBUS; his Life, Voyages, and Discoveries. Revised Edition, with 4 Woodcuts. 18mo. 2s. 6d.

CAPTAIN JAMES COOK; his Life, Voyages, and Discoveries. Revised Edition, with numerous Woodcuts. 18mo. 2s. 6d.

HUMBOLDT'S TRAVELS and DISCOVERIES in SOUTH AMERICA. Third Edition, with numerous Woodcuts. 18mo. 2s. 6d.

MUNGO PARK'S LIFE and TRAVELS in AFRICA, with an Account of his Death and the Substance of Later Discoveries. Sixth Edition, with Woodcuts. 18mo. 2s. 6d.

NARRATIVES of SHIPWRECKS of the ROYAL NAVY between 1793 and 1857, compiled from Official Documents in the Admiralty by W. O. S. GILLY; with a Preface by W. S. GILLY, D.D. Third Edition, fcp. 5s.

A WEEK at the LAND'S END. By J. T. BLIGHT; assisted by E. H. RODD, R. Q. COUCH, and J. RALFS. With Map and 96 Woodcuts. Fcp. price 6s. 6d.

VISITS to REMARKABLE PLACES: Old Halls, Battle-Fields, and Scenes Illustrative of Striking Passages in English History and Poetry. By WILLIAM HOWITT. 2 vols. square crown 8vo. with Wood Engravings, price 25s.

The RURAL LIFE of ENGLAND. By the same Author. With Woodcuts by Bewick and Williams. Medium 8vo. 12s. 6d.

Works of *Fiction*.

ATHERSTONE PRIORY. By L. N. COMYN. 2 vols. post 8vo. 21s.

Ellice: a Tale. By the same Author. Post 8vo. 9s. 6d.

STORIES and TALES by the Author of 'Amy Herbert,' uniform Edition, each Tale *or* Story complete in a single Volume.

AMY HERBERT, 2s. 6d.	IVORS, 3s. 6d.
GERTRUDE, 2s. 6d.	KATHARINE ASHTON, 3s. 6d.
EARL'S DAUGHTER, 2s. 6d.	MARGARET PERCIVAL, 5s.
EXPERIENCE OF LIFE, 2s. 6d.	LANETON PARSONAGE, 4s. 6d.
CLEVE HALL, 3s. 6d.	URSULA, 4s. 6d.

A Glimpse of the World. By the Author of 'Amy Herbert.' Fcp. 7s. 6d.

THE SIX SISTERS of the VALLEYS: an Historical Romance. By W. BRAMLEY-MOORE, M.A. Incumbent of Gerrard's Cross, Bucks. Third Edition, with 14 Illustrations. Crown 8vo. 5s.

The GLADIATORS: A Tale of Rome and Judæa. By G. J. WHYTE MELVILLE. Crown 8vo. 5s.

Digby Grand, an Autobiography. By the same Author. 1 vol. 5s.

Kate Coventry, an Autobiography. By the same. 1 vol. 5s.

General Bounce, or the Lady and the Locusts. By the same. 1 vol. 5s.

Holmby House, a Tale of Old Northamptonshire. 1 vol. 5s.

Good for Nothing, or All Down Hill. By the same. 1 vol. 6s.

The Queen's Maries, a Romance of Holyrood. 1 vol. 6s.

The Interpreter, a Tale of the War. By the same. 1 vol. 5s.

TALES from GREEK MYTHOLOGY. By George W. Cox, M.A. late Scholar of Trin. Coll. Oxon. Second Edition. Square 16mo. 3s. 6d.

Tales of the Gods and Heroes. By the same Author. Second Edition. Fcp. 5s.

Tales of Thebes and Argos. By the same Author. Fcp. 4s. 6d.

BECKER'S GALLUS; or, Roman Scenes of the Time of Augustus: with Notes and Excursuses illustrative of the Manners and Customs of the Ancient Romans. New Edition. [*Nearly ready.*

BECKER'S CHARICLES; a Tale illustrative of Private Life among the Ancient Greeks: with Notes and Excursuses. New Edition. [*Nearly ready.*

ICELANDIC LEGENDS. Collected by JON ARNASON. Selected and Translated from the Icelandic by G. E. J. POWELL and E. MAGNUSSON. SECOND SERIES, with Notes and an Introductory Essay on the Origin and Genius of the Icelandic Folk-Lore, and 3 Illustrations on Wood. Cr. 8vo. 21s.

The WARDEN: a Novel. By ANTHONY TROLLOPE. Crown 8vo. 3s. 6d.

Barchester Towers: a Sequel to 'The Warden.' By the same Author. Crown 8vo. 5s.

Poetry and The *Drama*.

GOETHE'S SECOND FAUST. Translated by JOHN ANSTER, LL.D. M.R.I.A. Regius Professor of Civil Law in the University of Dublin. Post 8vo. 15s.

TASSO'S JERUSALEM DELIVERED. Translated into English Verse by Sir J. Kingston James, Kt. M.A. 2 vols. fcp. with Facsimile, 14s.

POETICAL WORKS of JOHN EDMUND READE; with final Revision and Additions. 3 vols. fcp. 18s. or each vol. separately, 6s.

MOORE'S POETICAL WORKS, Cheapest Editions complete in 1 vol. including the Autobiographical Prefaces and Author's last Notes, which are still copyright. Crown 8vo. ruby type, with Portrait, 6s. or People's Edition, in larger type, 12s. 6d.

Moore's Poetical Works, as above, Library Edition, medium 8vo. with Portrait and Vignette, 14s. or in 10 vols. fcp. 3s. 6d. each.

MOORE'S IRISH MELODIES, 32mo. Portrait, 1s. 16mo. Vignette, 2s. 6d.

Maclise's Edition of Moore's Irish Melodies, with 161 Steel Plates from Original Drawings. Super-royal 8vo. 31s. 6d.

Maclise's Edition of Moore's Irish Melodies with all the Original Designs (as above) reduced by a New Process. Imp. 16mo. 10s. 6d.

MOORE'S LALLA ROOKH. 32mo. Plate, 1s. 16mo. Vignette, 2s. 6d.

Tenniel's Edition of Moore's Lalla Rookh, with 68 Wood Engravings from original Drawings and other Illustrations. Fcp. 4to. 21s.

SOUTHEY'S POETICAL WORKS, with the Author's last Corrections and copyright Additions. Library Edition, in 1 vol. medium 8vo. with Portrait and Vignette, 14s. or in 10 vols. fcp. 3s. 6d. each.

LAYS of ANCIENT ROME; with *Ivry* and the *Armada*. By the Right Hon. Lord Macaulay. 16mo. 4s. 6d.

Lord Macaulay's Lays of Ancient Rome. With 90 Illustrations on Wood, Original and from the Antique, from Drawings by G. Scharf. Fcp. 4to. 21s.

POEMS. By Jean Ingelow. Tenth Edition. Fcp. 8vo. 5s.

POETICAL WORKS of LETITIA ELIZABETH LANDON (L.E.L.) 2 vols. 16mo. 10s.

PLAYTIME with the POETS: a Selection of the best English Poetry for the use of Children. By a Lady. Revised Edition. Crown 8vo. 5s.

BOWDLER'S FAMILY SHAKSPEARE, cheaper Genuine Edition, complete in 1 vol. large type, with 36 Woodcut Illustrations, price 14s. or with the same Illustrations, in 6 pocket vols. 3s. 6d. each.

ARUNDINES CAMI, sive Musarum Cantabrigiensium Lusus canori. Collegit atque edidit H. Drury, M.A. Editio Sexta, curavit H. J. Hodgson, M.A. Crown 8vo. 7s. 6d.

The **ILIAD of HOMER TRANSLATED** into **BLANK VERSE.** By Ichabod Charles Wright, M.A. late Fellow of Magd. Coll. Oxon. 2 vols. crown 8vo. 21s.

The **ILIAD of HOMER in ENGLISH HEXAMETER VERSE.** By J. Henry Dart, M.A. of Exeter College, Oxford: Author of 'The Exile of St. Helena, Newdigate, 1838.' Square crown 8vo. 21s.

DANTE'S DIVINE COMEDY, translated in English Terza Rima by JOHN DAYMAN, M.A. [With the Italian Text, after *Brunetti*, interpaged.] 8vo. 21s.

Rural Sports, &c.

ENCYCLOPÆDIA of RURAL SPORTS; a complete Account, Historical, Practical, and Descriptive, of Hunting, Shooting, Fishing, Racing, &c. By D. P. BLAINE. With above 600 Woodcuts (20 from Designs by JOHN LEECH). 8vo. 42s.

NOTES on RIFLE SHOOTING. By Captain HEATON, Adjutant of the Third Manchester Rifle Volunteer Corps. Revised Edition. Fcp. 2s. 6d.

COL. HAWKER'S INSTRUCTIONS to YOUNG SPORTSMEN in all that relates to Guns and Shooting. Revised by the Author's SON. Square crown 8vo. with Illustrations, 18s.

The RIFLE, its THEORY and PRACTICE. By ARTHUR WALKER (79th Highlanders), Staff. Hythe and Fleetwood Schools of Musketry. Second Edition. Crown 8vo. with 125 Woodcuts, 5s.

The DEAD SHOT, or Sportsman's Complete Guide; a Treatise on the Use of the Gun, Dog-breaking, Pigeon-shooting, &c. By MARKSMAN. Revised Edition. Fcp. 8vo. with Plates, 5s.

HINTS on SHOOTING, FISHING, &c. both on Sea and Land and in the Fresh and Saltwater Lochs of Scotland; being the Experiences of C. IDLE. Second Edition, revised. Fcp. 6s.

The FLY-FISHER'S ENTOMOLOGY. By ALFRED RONALDS. With coloured Representations of the Natural and Artificial Insect. Sixth Edition; with 20 coloured Plates. 8vo. 14s.

HANDBOOK of ANGLING: Teaching Fly-fishing, Trolling, Bottom-fishing, Salmon-fishing; with the Natural History of River Fish, and the best modes of Catching them. By EPHEMERA. Fcp. Woodcuts, 5s.

The CRICKET FIELD; or, the History and the Science of the Game of Cricket. By JAMES PYCROFT, B.A. Fourth Edition. Fcp. 5s.

The Cricket Tutor; a Treatise exclusively Practical. By the same. 18mo. 1s.

Cricketana. By the same Author. With 7 Portraits. Fcp. 5s.

The HORSE-TRAINER'S and SPORTMAN'S GUIDE: with Considerations on the Duties of Grooms, on Purchasing Blood Stock, and on Veterinary Examination. By DIGBY COLLINS. Post 8vo. 6s.

The HORSE'S FOOT, and HOW to KEEP IT SOUND. By W. MILES, Esq. Ninth Edition, with Illustrations. Imperial 8vo. 12s. 6d.

A Plain Treatise on Horse-Shoeing. By the same Author. Post 8vo. with Illustrations, 2s. 6d.

Stables and Stable-Fittings. By the same. Imp. 8vo. with 13 Plates, 15s.

Remarks on Horses' Teeth, addressed to Purchasers. By the same. Post 8vo. 1s. 6d.

On DRILL and MANŒUVRES of CAVALRY, combined with Horse Artillery. By Major-Gen. MICHAEL W. SMITH, C.B. Commanding the Poonah Division of the Bombay Army. 8vo. 12s. 6d.

BLAINE'S VETERINARY ART: a Treatise on the Anatomy, Physiology, and Curative Treatment of the Diseases of the Horse, Neat Cattle and Sheep. Seventh Edition, revised and enlarged by C. STEEL, M.R.C.V.S.L. 8vo. with Plates and Woodcuts, 18s.

The HORSE: with a Treatise on Draught. By WILLIAM YOUATT. New Edition, revised and enlarged. 8vo. with numerous Woodcuts, 10s. 6d.

The Dog. By the same Author. 8vo. with numerous Woodcuts, 6s.

The DOG in HEALTH and DISEASE. By STONEHENGE. With 70 Wood Engravings. Square crown 8vo. 15s.

The Greyhound. By the same Author. Revised Edition, with 24 Portraits of Greyhounds. Square crown 8vo. 21s.

The OX; his Diseases and their Treatment: with an Essay on Parturition in the Cow. By J. R. DOBSON, M.R.C.V.S. Crown 8vo. with Illustrations. price 7s. 6d.

Commerce, Navigation, and Mercantile Affairs.

PRACTICAL GUIDE for BRITISH SHIPMASTERS to UNITED States Ports. By PIERREPONT EDWARDS, Her Britannic Majesty's Vice-Consul at New York. Post 8vo. 8s. 6d.

A NAUTICAL DICTIONARY, defining the Technical Language relative to the Building and Equipment of Sailing Vessels and Steamers, &c. By ARTHUR YOUNG. Second Edition; with Plates and 150 Woodcuts. 8vo. 18s.

A DICTIONARY, Practical, Theoretical, and Historical, of Commerce and Commercial Navigation. By J. R. M'CULLOCH, Esq. 8vo. with Maps and Plans, 50s.

A MANUAL for NAVAL CADETS. By J. M'NEIL BOYD, late Captain R.N. Third Edition; with 240 Woodcuts and 11 coloured Plates. Post 8vo. 12s. 6d.

The LAW of NATIONS Considered as Independent Political Communities. By TRAVERS TWISS, D.C.L. Regius Professor of Civil Law in the University of Oxford. 2 vols. 8vo. 30s. or separately, PART I. Peace, 12s. PART II. War, 18s.

Works of Utility and General Information.

MODERN COOKERY for PRIVATE FAMILIES, reduced to a System of Easy Practice in a Series of carefully-tested Receipts. By ELIZA ACTON. Newly revised and enlarged; with 8 Plates, Figures, and 150 Woodcuts. Fcp. 7s. 6d.

The HANDBOOK of DINING; or. Corpulency and Leanness scientifically considered. By BRILLAT-SAVARIN, Author of 'Physiologie du Goût.' Translated by L. F. SIMPSON. Revised Edition, with Additions. Fcp. 3s. 6d.

On FOOD and its DIGESTION; an Introduction to Dietetics. By W. BRINTON, M.D. Physician to St. Thomas's Hospital, &c. With 48 Woodcuts. Post 8vo. 12s.

WINE, the VINE, and the CELLAR. By Thomas G. Shaw. Second Edition, revised and enlarged, with Frontispiece and 31 Illustrations on Wood. 8vo. 16s.

HOW TO BREW GOOD BEER: a complete Guide to the Art of Brewing Ale, Bitter Ale, Table Ale, Brown Stout, Porter, and Table Beer. By John Pitt. Revised Edition. Fcp. 4s. 6d.

A PRACTICAL TREATISE on BREWING; with Formulæ for Public Brewers, and Instructions for Private Families. By W. Black. 8vo. 10s. 6d.

SHORT WHIST. By Major A. Sixteenth Edition, revised, with an Essay on the Theory of the Modern Scientific Game by Prof. P. Fcp. 3s. 6d.

WHIST, WHAT TO LEAD. By Cam. Third Edition. 32mo. 1s.

HINTS on ETIQUETTE and the USAGES of SOCIETY; with a Glance at Bad Habits. Revised, with Additions, by a Lady of Rank. Fcp. price 2s. 6d.

TWO HUNDRED CHESS PROBLEMS, composed by F. Healey, including the Problems to which the Prizes were awarded by the Committees of the Era, the Manchester, the Birmingham, and the Bristol Chess Problem Tournaments; accompanied by the Solutions. Crown 8vo. with 200 Diagrams, 5s.

The CABINET LAWYER; a Popular Digest of the Laws of England, Civil and Criminal. Twenty-first Edition, extended by the Author; including the Acts of the Sessions 1864 and 1865. Fcp. 10s. 6d.

The PHILOSOPHY of HEALTH; or, an Exposition of the Physiological and Sanitary Conditions conducive to Human Longevity and Happiness. By Southwood Smith, M.D. Eleventh Edition, revised and enlarged: with 113 Woodcuts. 8vo. 15s.

HINTS to MOTHERS on the MANAGEMENT of their HEALTH during the Period of Pregnancy and in the Lying-in Room. By T. Bull, M.D. Fcp. 5s.

The Maternal Management of Children in Health and Disease. By the same Author. Fcp. 5s.

The LAW RELATING to BENEFIT BUILDING SOCIETIES; with Practical Observations on the Act and all the Cases decided thereon; also a Form of Rules and Forms of Mortgages. By W. Tidd Pratt, Barrister. Second Edition. Fcp. 3s. 6d.

NOTES on HOSPITALS. By Florence Nightingale. Third Edition, enlarged; with 13 Plans. Post 4to. 18s.

C. M. WILLICH'S POPULAR TABLES for ascertaining the Value of Lifehold, Leasehold, and Church Property, Renewal Fines, &c.: the Public Funds; Annual Average Price and Interest on Consols from 1731 to 1861; Chemical, Geographical, Astronomical, Trigonometrical Tables, &c. Post 8vo. 10s.

THOMSON'S TABLES of INTEREST, at Three, Four, Four and a Half, and Five per Cent. from One Pound to Ten Thousand and from 1 to 365 Days. 12mo. 3s. 6d.

MAUNDER'S TREASURY of KNOWLEDGE and LIBRARY of Reference: comprising an English Dictionary and Grammar, Universal Gazetteer, Classical Dictionary, Chronology, Law Dictionary, a Synopsis of the Peerage, useful Tables, &c. Fcp. 10s.

INDEX.

Abbott on Sight and Touch 10
Acton's Modern Cookery 27
Alcock's Residence in Japan 22
Allies on Formation of Christendom 20
Alpine Guide (The) 22
Apjohn's Manual of the Metalloids 12
Arago's Biographies of Scientific Men 5
———— Popular Astronomy 10
Arnold's Manual of English Literature 7
Arnott's Elements of Physics 11
Arundines Cami 25
Atherstone Priory 23
Autumn holidays of a Country Parson .. 8
Ayre's Treasury of Bible Knowledge 19

Bacon's Essays, by Whately 5
———— Life and Letters, by Spedding 5
———— Works 6
Bain on the Emotions and Will 10
———— on the Senses and Intellect 10
———— on the Study of Character 10
Baines's Explorations in S. W. Africa 22
Ball's Alpine Guide 23
Barnard's Drawing from Nature 16
Baylon's Rents and Tillages 18
Beaten Tracks 22
Becker's Charicles and Gallus 24
Beethoven's Letters 4
Benfey's Sanskrit Dictionary 8
Berry's Journals and Correspondence 4
Black's Treatise on Brewing 28
Blackley and Friedländer's German and English Dictionary 8
Blaine's Rural Sports 26
———— Veterinary Art 27
Blight's Week at the Land's End 23
Boase's Essay on Human Nature 9
———— Philosophy of Nature 9
Booth's Epigrams 9
Boner's Transylvania 22
Bonney's Alps of Dauphiné 22
Bourne on Screw Propeller 17
Bourne's Catechism of the Steam Engine .. 17
———— Handbook of Steam Engine 17
———— Treatise on the Steam Engine 17
Bowdler's Family Shakspeare 25
Boyd's Manual for Naval Cadets 27
Bramley-Moore's Six Sisters of the Valleys 24
Brande's Dictionary of Science, Literature, and Art 13
Bray's (C.) Education of the Feelings 10
———— Philosophy of Necessity 10
Brinton on Food and Digestion 27
Bristow's Glossary of Mineralogy 11
Brodie's (Sir C. B.) Works 15
———— Autobiography 15
———— Constitutional History 2

Browne's Ice Caves of France and Switzerland 15
———— Exposition 39 Articles 18
———— Pentateuch 18
Buckle's History of Civilization 2
Bull's Hints to Mothers 28
———— Maternal Management of Children 28
Bunsen's Ancient Egypt 3
Bunsen on Apocrypha 20
Burke's Vicissitudes of Families 5
Burton's Christian Church 3

Cabinet Lawyer 28
Calvert's Wife's Manual 21
Campaigner at Home 8
Cats and Farlie's Moral Emblems 16
Chorale Book for England 22
Clough's Lives from Plutarch 2
Colenso (Bishop) on Pentateuch and Book of Joshua 19
Collins's Horse-Trainer's Guide 26
Columbus's Voyages 23
Commonplace Philosopher in Town and Country 8
Conington's Handbook of Chemical Analysis 13
Contanseau's Pocket French and English Dictionary 8
———— Practical ditto 8
Conybeare and Howson's Life and Epistles of St. Paul 18
Cook's Voyages 23
Copland's Dictionary of Practical Medicine 15
———— Abridgment of ditto 15
Cox's Tales of the Great Persian War 2
———— Tales from Greek Mythology 24
———— Tales of the Gods and Heroes 24
———— Tales of Thebes and Argos 24
Cresy's Encyclopædia of Civil Engineering 17
Critical Essays of a Country Parson 8
Crowe's History of France 2
Cussans's Grammar of Heraldry 16

Dart's Iliad of Homer 25
Dayman's Dante's Divina Commedia 26
D'Aubigné's History of the Reformation in the time of Calvin 2
Dead Shot (The), by Marksman 26
De la Rive's Treatise on Electricity 11
Delmard's Village Life in Switzerland 22
De la Pryme's Life of Christ 20
De Morgan on Matter and Spirit 9
De Tocqueville's Democracy in America .. 2
Dobson on the Ox 27
Duncan and Millard on Classification, &c. of the Idiotic 15
Dyer's City of Rome 2

Edwards' Shipmaster's Guide 27
Elements of Botany 15
Ellice, a Tale 23
Ellicott's Broad and Narrow Way 19
——— Commentary on Ephesians 19
——— Destiny of the Creature........ 19
——— Lectures on Life of Christ 19
——— Commentary on Galatians 19
——————— Pastoral Epist.... 19
——————————— Philippians, &c... 19
——————————— Thessalonians.... 19
Essays and Reviews 20
——— on Religion and Literature, edited by Manning 20

Fairbairn on Iron Shipbuilding 17
Fairbairn's Application of Cast and Wrought Iron to Building............... 17
——— Information for Engineers... 17
——— Treatise on Mills & Millwork 17
Farrar's Chapters on Language 7
Foulkes's Christendom's Divisions........ 20
Freshfield's Alpine Byways 23
——— Tour in the Grisons 23
Friends in Council 9
Froude's History of England 1

Garratt's Marvels and Mysteries of Instinct 12
Gee's Sunday to Sunday................... 21
Gilbert and Churchill's Dolomite Mountains 22
Gilly's Shipwrecks of the Navy 23
Goethe's Second Faust, by Anster 24
Goodeve's Elements of Mechanism 17
Gorle's Questions on Browne's Exposition of the 39 Articles...................... 18
Grant's Ethics of Aristotle 5
Graver Thoughts of a Country Parson 8
Gray's Anatomy 14
Greene's Corals and Sea Jellies 12
——— Sponges and Animalculæ 12
Grove on Correlation of Physical Forces.. 11
Gwilt's Encyclopædia of Architecture 16

Handbook of Angling, by Ephemera 26
Hare on Election of Representatives 6
Hartwig's Sea and its Living Wonders.... 12
——— Harmonies of Nature 12
——— Tropical World 12
Haughton's Manual of Geology........... 11
Hawker's Instructions to Young Sportsmen 26
Healey's Chess Problems 28
Heaton's Notes on Rifle Shooting 26
Helps's Spanish Conquest in America..... 2
Herschel's Essays from the Edinburgh and Quarterly Reviews 13
——— Outlines of Astronomy 10
Hewitt on the Diseases of Women 14
Hints on Etiquette........................ 28
Hodgson's Time and Space............... 10
Holland's Essays on Scientific Subjects... 13
Holmes's System of Surgery.............. 14
Hooker and Walker-Arnott's British Flora.................................. 13
Horne's Introduction to the Scriptures 19
——— Compendium of ditto......... 19
Horsley's Manual of Poisons 15
Hoskyns's Talpa.......................... 18
How we Spent the Summer 22
Howitt's Australian Discovery 22
——— Rural Life of England....... 25
——— Visits to Remarkable Places.. 23
Howson's Hulsean Lectures on St. Paul... 18

Hughes's (W.) Geography of British History 11
——— Manual of Geography 11
Hullah's History of Modern Music........ 4
——— Transition Musical Lectures 4
Humboldt's Travels in South America ... 23
Humphreys' Sentiments of Shakspeare ... 16
Hutton's Studies in Parliament 9
Hymns from Lyra Germanica............. 21

Icelandic Legends. Second Series 24
Idle's Hints on Shooting 26
Ingelow's Poems.......................... 25

Jameson's Legends of the Saints and Martyrs...................................... 16
——— Legends of the Madonna....... 16
——— Legends of the Monastic Orders 16
Jameson and Eastlake's History of Our Lord 16
John's Home Walks and Holiday Rambles 12
Johnson's Patentee's Manual 17
——— Practical Draughtsman...... 17
Johnston's Gazetteer, or Geographical Dictionary 11
Jones's Christianity and Common Sense.... 10

Kalisch's Commentary on the Bible 7
——— Hebrew Grammar............. 7
Kesteven's Domestic Medicine 15
Kirby and Spence's Entomology 12
Kuenen on Pentateuch and Joshua 19

Lady's Tour Round Monte Rosa 23
Landon's (L. E. L.) Poetical Works....... 25
Latham's English Dictionary 7
Lecky's History of Rationalism 3
Leisure Hours in Town 8
Lewes' History of Philosophy 3
Lewin's Fasti Sacri 19
Lewis on Early Roman History 6
——— Essays on Administrations 6
——— Fables of Babrius............. 6
——— on Foreign Jurisdiction 6
——— on Irish Disturbances 6
——— on Observation and Reasoning in Politics................................. 6
——— on Political Terms 6
Liddell and Scott's Greek-English Lexicon 8
——— Abridged ditto 8
Life of Man Symbolized 16
Lindley and Moore's Treasury of Botany 12
Longman's Lectures on the History of England 2
Loudon's Agriculture 18
——— Cottage, Farm, Villa Architecture 18
——— Gardening 18
——— Plants........................ 13
——— Trees and Shrubs 13
Lowndes' Engineer's Handbook 16
Lyra Domestica 21
——— Eucharistica 21
——— Germanica 16, 21
——— Messianica 21
——— Mystica...................... 21
——— Sacra 21

Macaulay's (Lord) Essays 3
——— History of England 1
——— Lays of Ancient Rome....... 25
——— Miscellaneous Writings 9
——— Speeches 7
——— Works 1

MACDOUGALL's Theory of War.................. 17
McCULLOCH's Dictionary of Commerce 27
——————— Geographical Dictionary....... 11
MACFIE's Vancouver Island 22
MAGUIRE's Life of Father Mathew........... 4
——————— Rome and its Rulers............ 4
MALING's Indoor Gardener 13
MANNING on Holy Ghost........................ 20
MARSHMAN's Life of Havelock................ 5
MASSEY's History of England................ 1
MASSINGBERD's History of the Reformation.. 4
MAUNDER's Biographical Treasury 5
——————— Geographical Treasury 11
——————— Historical Treasury 3
——————— Scientific and Literary Treasury 13
——————— Treasury of Knowledge 28
——————— Treasury of Natural History .. 12
MAURY's Physical Geography 10
MAY's Constitutional History of England... 1
MELVILLE's Digby Grand....................... 24
——————— General Bounce 24
——————— Gladiators 24
——————— Good for Nothing 24
——————— Holmby House 24
——————— Interpreter 24
——————— Kate Coventry 24
——————— Queen's Maries................ 24
MENDELSSOHN's Letters........................ 4
MENZIES' Windsor Great Park............... 18
MERIVALE's (H.) Historical Studies 2
——————— (C.) Fall of the Roman Republic 3
——————— Boyle Lectures 3
——————— Romans under the Empire 2
MILES on Horse's Foot and Horseshoeing... 26
——————— on Horses' Teeth and Stables..... 26
MILL on Liberty.................................. 6
——————— on Representative Government .. 6
——————— on Utilitarianism.................. 6
MILL's Dissertations and Discussions 6
——————— Political Economy 6
——————— System of Logic 6
——————— Hamilton's Philosophy 6
MILLER's Elements of Chemistry 14
MONSELL's Spiritual Songs 21
——————— Beatitudes 21
MONTGOMERY on Pregnancy ....
MOORE's Irish Melodies........................ 25
——————— Lalla Rookh 25
——————— Journal and Correspondence ... 5
——————— Poetical Works 25
MORELL's Elements of Psychology 9
——————— Mental Philosophy 9
Morning Clouds 20
MOSHEIM's Ecclesiastical History 20
MOZART's Letters.............................. 4
MÜLLER's (Max) Lectures on the Science of Language 7
——————— (K. O.) Literature of Ancient Greece 2
MURCHISON on Continued Fevers........... 14
MURE's Language and Literature of Greece 2

New Testament, Illustrated with Wood Engravings from the Old Masters............ 15
NEWMAN's History of his Religious Opinions 4
NIGHTINGALE's Notes on Hospitals........... 28

ODLING's Animal Chemistry 14
——————— Course of Practical Chemistry.... 14
——————— Manual of Chemistry 14
ORMSBY's Rambles in Algeria and Tunis... 22
O'SHEA's Guide to Spain 23
OWEN's Comparative Anatomy and Physiology of Vertebrate Animals 12
OXENHAM on Atonement...................... 21

PACKE's Guide to the Pyrenees 23
PAGET's Lectures on Surgical Pathology .. 14
PARK's Life and Travels...................... 23
PEREIRA's Elements of Materia Medica.... 15
——————— Manual of Materia Medica 15
PERKINS's Tuscan Sculptors 16
PHILLIPS's Guide to Geology 11
——————— Introduction to Mineralogy 11
PIESSE's Art of Perfumery 28
——————— Chemical, Natural, and Physical Magic 18
PITT on Brewing............................... 24
Playtime with the Poets 25
Practical Mechanic's Journal 17
PRATT's Law of Building Societies 28
PRESCOTT's Scripture Difficulties 19
PROCTOR's Saturn 10
PYCROFT's Course of English Reading...... 7
——————— Cricket Field 26
——————— Cricket Tutor 26
——————— Cricketana 26

READE's Poetical Works 25
Recreations of a Country Parson............. 8
REILY's Map of Mont Blanc 22
RIDDLE's First Sundays at Church 21
RIVERS's Rose Amateur's Guide............ 13
ROGERS's Correspondence of Greyson 9
——————— Eclipse of Faith 9
——————— Defence of ditto 9
——————— Essays from the Edinburgh Review 9
——————— Fulleriana 9
ROGET's Thesaurus of English Words and Phrases 7
RONALDS's Fly-Fisher's Entomology 26
ROWTON's Debater............................. 7
RUSSELL on Government and Constitution.. 1

SANDARS's Justinian's Institutes............ 5
SCOTT's Handbook of Volumetrical Analysis 14
SCROPE on Volcanos 11
SENIOR's Essays................................ 3
SEWELL's Amy Herbert 24
——————— Cleve Hall....................... 24
——————— Earl's Daughter................. 24
——————— Examination for Confirmation ... 20
——————— Experience of Life.............. 24
——————— Gertrude 24
——————— Glimpse of the World 24
——————— History of the Early Church... 3
——————— Ivors............................. 24
——————— Katharine Ashton 24
——————— Laneton Parsonage 24
——————— Margaret Percival 24
——————— Night Lessons from Scripture 20
——————— Passing Thoughts on Religion... 20
——————— Preparation for Communion..... 20
——————— Principles of Education 20
——————— Readings for Confirmation..... 20
——————— Readings for Lent 20
——————— Stories and Tales 24
——————— Thoughts for the Holy Week 20
——————— Ursula 24
SHAW's Work on Wine 28
SHEDDEN's Elements of Logic 6
SHIPLEY's Church and the World 19
Short Whist 28
SHORT's Church History 3
SIEVEKING's (AMELIA) Life, by WINKWORTH 4
SIMPSON's Handbook of Dining 27
SMITH's (SOUTHWOOD) Philosophy of Health 28
——————— (J.) Paul's Voyage and Shipwreck.. 18
——————— (G.) Wesleyan Methodism 4
——————— (SYDNEY) Memoir and Letters.. 5
——————— Miscellaneous Works 9
——————— Moral Philosophy 9
——————— Wit and Wisdom.......... 9

SMITH on Cavalry Drill and Manœuvres	26
SOUTHEY's (Doctor)	7
———— Poetical Works	25
STANLEY's History of British Birds	12
STEBBING's Analysis of MILL's Logic	6
STEPHEN's Essays in Ecclesiastical Biography	5
———— Lectures on History of France	2
STIRLING's Secret of Hegel	10
STONEHENGE on the Dog	27
———— on the Greyhound	27
STRANGE on Sea Air	15
———— on Restoration of Health	15
TASSO's Jerusalem, by JAMES	25
TAYLOR's (Jeremy) Works, edited by EDEN	20
TENNENT's Ceylon	12
———— Natural History of Ceylon	12
THIRLWALL's History of Greece	2
THOMSON's (Archbishop) Laws of Thought	6
———— (J.) Tables of Interest	28
———— Conspectus, by BIRKETT	15
TODD's Cyclopædia of Anatomy and Physiology	14
———— and BOWMAN's Anatomy and Physiology of Man	15
TROLLOPE's Barchester Towers	24
———— Warden	24
TWISS's Law of Nations	27
TYNDALL's Lectures on Heat	11
URE's Dictionary of Arts, Manufactures, and Mines	17
VAN DER HOEVEN's Handbook of Zoology	12
VAUGHAN's (R.) Revolutions in English History	1
———— Way to Rest	10
———— (R. A.) Hours with the Mystics	10

WALKER on the Rifle	26
WATSON's Principles and Practice of Physic	14
WATTS's Dictionary of Chemistry	13
WEBB's Objects for Common Telescopes	10
WEBSTER & WILKINSON's Greek Testament	19
WELD's Last Winter in Rome	22
WELLINGTON's Life, by BRIALMONT and GLEIG	4
———— by GLEIG	4
WEST on Children's Diseases	14
WHATELY's English Synonymes	5
———— Logic	5
———— Remains	6
———— Rhetoric	5
———— Sermons	21
———— Paley's Moral Philosophy	21
WHEWELL's History of the Inductive Sciences	3
———— Scientific Ideas	3
WHIST, what to lead, by CAM	28
WHITE and RIDDLE's Latin-English Dictionaries	7
WILBERFORCE (W.) Recollections of, by HARFORD	5
WILLICH's Popular Tables	28
WILSON's Bryologia Britannica	13
WINDHAM's Diary	4
WOOD's Homes without Hands	12
WOODWARD's Historical and Chronological Encyclopædia	3
WRIGHT's Homer's Iliad	25
YONGE's English-Greek Lexicon	8
———— Abridged ditto	8
YOUNG's Nautical Dictionary	27
YOUATT on the Dog	27
———— on the Horse	27

www.ingramcontent.com/pod-product-compliance
Lightning Source LLC
Chambersburg PA
CBHW030014240426
43672CB00007B/949